From Salem to Nashville

OLD GLORY

The Life and Times of Patriot Captain William Driver

Delivered by Great-Great-Grandson—Jack Benz

Co-author—Garrett Williams

Acknowledgments

Jack Benz *Author and Publisher*
Garrett Williams *Co-author and Editor*
Nancy Adams Arnold *Graphic Designer and Production*

"Although my sixty-year collection of Driver history and memorabilia are unique resources, my limited experience in academic writing prompted me to solicit the assistance of recognized professionals. It is with special gratitude that I highlight the exceptional contributions of my co-author and graphic designer. Refer to page 273. Without their depth of knowledge and unique skills, this book never would have been written.
Gratefully, Jack Benz."

First Edition, 2019

Printed and bound in the United States of America

Available through amazon.com and bookstores

ISBN: 978-0-692-17556-9 (Paperback)
ISBN: 978-0-692-17583-5 (eBook)

Library of Congress Control Number: 2019913387

For permission requests and inquiries, visit our website at www.captaindriver.com

COVER PAGE CREDITS
Flag photo by Nancy Arnold. Driver Monument photo by Garrett Williams

Special recognition and appreciation are offered to:

Proofreaders: *Walker Batts, Noroma Benz, Harry Chapman, Charles Overby, Pam Garrett, Nancy Toon, and Jack Van Hooser*

Glyn Richard Acraman Web-master, Acraman Family Website, MyHeritage.

Billy Benz, my brother, who joins me as the only other living witness having heard Grandmother Georgie Wade's accounts that came directly from her Grandfather, Captain William Driver.

Fletch Coke for her research on Christ Church, Episcopal (Driver's church).

Dick Elliott for his original sailing vessel painting depicted on page 1.

Herbert Ford, Director of the Pitcairn Islands Study Center for his Consultant services.

Carolyn Gregory for interpreting and typing Driver's handwritten letters.

J. Mark Lowe, Genealogist, Family Historian.

Stephen and Beverly Mansfield, Authors, for encouraging me to undertake and complete this project.

Bill Mckee for his guidance and research.

Bill R. Phillips for his assistance in our research.

Dan Pomeroy, Director of the Tennessee State Museum, for his general research.

Merryanne Pierson, Genealogist, for her research on the Driver family.

Dick Waggener, Historian, First Baptist Church, Nashville, church research.

Index of Resource Authors

Rupert S. Holland *Yankee Ships in Pirate Waters—Children of the Son*

Louise Davis *Frontier Tales of Tennessee—Captain Driver and the flag—et.al*

Captain Henry Nichols *Eastward Around the World on the Emerald*

Kenneth Davis *America's Hidden History*

Jan Prince *Tahiti*

Dick Schaap *A Bridge to the Seven Seas*

Geo Tindall & Dave Shi *America—A Narrative History*

Garvin Menzies *Who Discovered America and 1421: The Year China Discovered America*

John Merrill *Old Glory Driver*

Robert Guron *The Life and Achievements of Old Glory Driver*

Rear-Admiral Wm Furlong, CMD Byron McCandless, and Harold Langley *So Proudly we Hail: The History of the United States Flag*

Mary Driver Roland *Old Glory: The True Story*

Harriet Ruth Cooke *The Driver Descendants of Robert and Phebe Driver*

Edgar Maclay *A History of American Privateers*

Dan Pomeroy *His Ship, His Country, and His Flag*

Ancestry.com Fold3 Newspapers.com

Smithsonian Institution Archives

Australian National University—Pacific Manuscript Bureau

American Merchant Marine at War—www.usmm.org

Phillips Library at Peabody Essex Museum

Tales from my Grandmother—Georgie Wade

Dedication

This book is dedicated in memory of my Grandmother,

Georgie Driver Reece Wade,

who is the most credible witness of all
in the coming account about the life and times
of Captain William Driver.

She was his granddaughter and was 13 years old
at the time of his death in 1886. Not only did she and her
mother live in the same city as he, Nashville, Tennessee,
she was also old enough to recall having directly
heard him share his captivating stories.

As a young boy, I frequently listened to
Grandmother Georgie recount many of my
great-great-grandfather's exciting sea adventures
around the world. I owe her my eternal gratitude
for planting the urge within me to sustain his memory
and serve his mission to encourage all citizens
to honor and respect the flag of the
United States of America.

"O Captain! my Captain! our fearful trip is done;
the ship has weather'd every rack, the prize we sought is won;
The port is near, the bells I hear, the people all exulting."[1]

Oh Captain! My Captain!—Walt Whitman

Painting by artist, Dick Elliott, in honor of Captain William Driver and this book.[1]

Foreword

By Jack Van Hooser

It has been my special privilege to have known Jack Benz for over half a century; we were classmates at Tennessee Tech University in the 1950s. For prudent reasons, he and I became the logical choices to be our family historians and have shared our stories and mutual interest for decades. Accordingly, I urged him to write a book about his very distinguished great-great-grandfather. I am thrilled in this foreword to affirm that the book is now a reality.

Jack Benz has penned what I believe is the definitive account of one of America's greatest patriots and historically significant achievers—Captain William Driver of Salem, Massachusetts, ports around the world, and Nashville, Tennessee.

Jack cleverly tells his story by embodying the persona of Captain Driver and re-engaging with the people, places, and times where the saga took place. Bon Voyage!

The Author's Message to the Readers

The life and times of Captain William Driver have been presented in numerous formats since his death in 1886. They have been conveyed in anthologies, books, short stories, magazine articles, newspaper accounts, and verbally from a variety of platforms and podiums. Most versions have followed a similar pattern of delivery. They tend to be general summaries about his life and travels but with major emphasis on two events, the enshrinement of his famous flag, *Old Glory*, and his involvement in what indirectly became an Academy Award Movie, *The Mutiny on the Bounty*. You will find this narrative to be different from all its predecessors in several respects and for significant reasons.

Foremost, I was born and grew up in Nashville, Tennessee, where William Driver, my Great-Great-Grandfather, lived most of his life. Consequently, special circumstances and unique coincidences positioned me with proprietary resources and direct experiences which are incomparable. For over sixty years, I collected and inherited a massive amount of Captain Driver's personal documents, records, and memorabilia. This includes his Bible containing extensive family records, large numbers of personal letters written in his own hand, microfilm of his 200-year-old ship's log, and much of what has been published about him through his entire life.

Additionally, to my knowledge, I am the only survivor carrying in my memory clear recall of hundreds of accounts, adventures, and stories told directly by Captain Driver to his Granddaughter and my Grandmother, Georgie Wade. These same captivating tales continually were repeated to me during my countless visits to her house throughout my childhood. Unfortunately, the memories of others who may have been present have faded, or they have passed away taking their recollections with them. It remains that I alone have retained many of the interesting details and pertinent facts that provide the most complete accounting of William Driver's life and thought; being in my eighties, I feel an urgency to get them on the record.

When I was a young child, Captain Driver became my alter ego in my dreams and make-believe games. I too had visions of sailing the "Seven Seas" to far away places with strange-sounding names.[2] In my mind's eye, I saw and imagined things about which you will read—amazing giants, frightening cannibals, death-defying storms, and circling the earth on a ship.

My fascination continued as a young adult when my brother and I purchased a sailboat and became real sailors on a hometown lake. My zeal strengthened with age as I became a competitive enthusiast on many of the lakes throughout the United States—winning several major championships.

Therefore, in the context of the foregoing, this narrative is intended to fill in the gaps and add missing segments appropriate to complete this incredible story about one of America's most unforgettable characters and staunchest patriots. I write this account in the persona of Captain Driver combining undisputed facts, hearsay, circumstantial evidence, and inferences to make assumptions about things that likely and logically took place but heretofore have been unaddressed.

At its core, this book is about patriotism. He was guided by values and principles written by our founding fathers in documents fittingly known as the foundation documents. They are included in the appendix of this book for the convenience of those who desire to reference them again. Those unfamiliar with the wisdom and entitlements contained therein will learn why the United States—in less than two and one-half centuries—has become the most advanced, influential, and generous country the world has ever known. I hope that Captain William Driver's story will reawaken the "Spirit of '76."

If you find this version to be clarifying, instructive, entertaining, inspiring, thought-provoking, and refreshingly different, my mission will have been accomplished. *Jack Benz-2019*

Table of Contents

"SOME YEARS AGO...HAVING LITTLE OR NO MONEY IN MY PURSE, AND NOTHING PARTICULAR TO INTEREST ME ON SHORE, I THOUGHT I WOULD SAIL ABOUT A LITTLE AND SEE THE WATERY PART OF THE WORLD... THERE IS NOTHING SURPRISING IN THIS...IF THEY BUT KNEW IT, ALMOST ALL MEN IN THEIR DEGREE, SOME TIME OR OTHER, CHERISH VERY NEARLY THE SAME FEELINGS TOWARDS THE OCEAN WITH ME."[3]

Loomings of Moby Dick, by Herman Melville

Red Sun [2]

Prologue

In a time that was measureless, strangely without a beginning or end, God moved throughout a shapeless expanse and creation emerged—first on the face of the waters, then the firmament, and lastly on the land.[4]

Evidence that our creator is omnipotent, omnipresent, and omniscient is compelling even to the weakest eyes and minds; yet, much is not so obvious. There are baffling intricacies, functions, and applications hidden from common view in the realm of the unknown and not understood. However, revelations patiently wait—often for decades—for a curious explorer to come and discover their secrets and master their processes.[5]

ESA/Hubble and NASA Supernova in Galaxy in NGC 3810 [3]

Most pioneers search, study, and live in one domain, but the trade of a mariner requires knowledge and collaboration of three—the water, sky, and land. Regardless, the sea is still the mariner's home.

It is a paradox that the vast oceans occupy nearly three-fourths of the earth's surface but do not host three-fourths of its people. More so, it is not ordinary folks who take to the sea in ships. They are unusual souls not daunted by the unknown or unthinkable. They fearlessly take risks, dodge hazards, endure hardships, and even chance death, for adventure, noble causes, and the thrill of the pursuit. We owe much to these daring discoverers and messengers seemingly destined to enlighten the world.

Indeed, they are special people—those restless adventurers who fill the voids left by the timid and inept. Although their marks are mere specks in this marvelous creation, their contributions are measureless.

A Ship in Need in a Raging Storm [4]

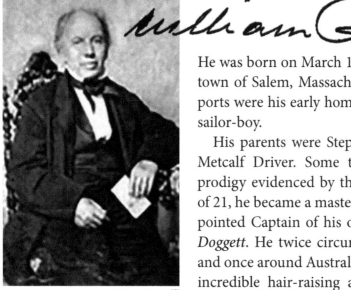

Captain William Driver [5]

He was born on March 17, 1803, in the harbor town of Salem, Massachusetts. Ships and seaports were his early homes starting as a young sailor-boy.

His parents were Stephen Driver and Ruth Metcalf Driver. Some thought of him as a prodigy evidenced by the fact that by the age of 21, he became a master mariner and was appointed Captain of his own ship, *The Charles Doggett*. He twice circumnavigated the world and once around Australia, experiencing many incredible hair-raising adventures. He wrote about them in his letters and logs.

One significant voyage to the South Pacific resulted in an exploit that produced an episode in the award-winning movie, "The Mutiny on the Bounty." However, his biggest claim to fame involved his famous flag, *Old Glory*, being enshrined in the Smithsonian Institution in Washington, D. C.

When the Union Army prevailed over the Confederates in Nashville during the Civil War, he risked his life by hoisting his personal Union flag to replace the Confederate flag on the State Capital of Tennessee. He was well aware that it was highly dangerous and politically incorrect at that time and place, and was widely renounced for doing it, even by three of his sons who became Confederate soldiers. Nevertheless, he fearlessly defended the sanctity and honor of the flag of the Union—the official flag of the United States of America.

Massachusetts and Tennessee state outlines and flags [6]

He became a man of unwavering faith in God and a patriot who practiced "kneeling before his maker and standing before his flag." He trusted that his example would offer a perspective about how the American flag came to be, what it represents, and why it is considered to be the most hallowed icon for freedom-loving patriots—as relevant in the future as it had been throughout his life. By his actions, he demonstrated why he chose to risk death rather than disrespect it.

He boldly flew his personal Union flag to assert that it is the American citizen's most precious symbol as their guardian of life, liberty, and the pursuit of happiness. It is the unconditional sentinel and friend of every race, color, and creed—bought and sustained by the blood of many thousands since the day of The American Revolution. In William Driver's value system, dishonoring the American flag is an incomparable act of defiance—fully equal to heresy and treason.

Ironically, though he believed such behavior was despicable and irrational, it is because of the flag's symbolic virtues that a dissenter is guaranteed the right to disrespect it. Even in the face of insults, the flag continues to fly courageously as the

A mysterious visitor, Rodney Acraman (insert at the top left) Captain William Driver (insert at top right) & Great-Great-Grandson Jack Benz at Captain Driver Tomb and Monument [7]

official protestor against all unrighteous acts, words, and deeds—all the bad, evil, inhumanity, and unfairness that misguided minds can conceive. It is not plausible, sensible, or evidence of clear understanding that one would protest against the very "Mother of Protest."

An attempt to parse words and put a noble face on repulsive behavior is further evidence of shallow reasoning. It simply is abnormal to attack one's defender, so to explain it any other way is an unconvincing deflection from the obvious facts. The mission achieves only the opposite of its intended purpose; it is a path doomed to failure along with all who follow it. Punishment, ridicule, resentment, and wrath are the wages for mistakes in common sense; and prevailing public opinion will enforce the penalties.

Society, in general, has failed to present a convincing rationale on behalf of patriotism. The nation's leaders, civic clubs, classrooms, and public forums must refocus on what easily should be the most compelling message ever delivered. Let us stand to honor our icon's presence—the symbol assuring liberty and justice for all in America.

> "There are hermit souls that live withdrawn in the peace of their self-content. There are souls, like stars, that dwell apart in a fellowless firmament. There are pioneer souls that blaze their paths where highways never ran . . ." [6]
>
> *The House by the Side of the Road*
> - Sam Walter Foss

William Driver's last voyage began in Nashville, Tennessee on the 2nd day of March, 1886—leaving indelible imprints around the world. His remains are at rest in the Nashville City Cemetery.

William Driver's Frame of Reference

To fully understand William Driver's story, one must learn about his roots. His forefathers were among America's earliest European settlers but in no way were the first Pilgrims. The setting was well-established. Studies by anthropologists, archeologists, and other specialists agree that the continent has been inhabited for thousands of years, but they are not in agreement as to the chronology—which came first.

"DNA testing and radiocarbon dating have confirmed that many of the earliest immigrants came from Siberia to Alaska via a land-bridge—Beringia—that crossed the Bering Strait."[7] "Others are thought to have come by boat from the same region as evidenced by 13,000-year-old human bones found on the Channel Islands off the California coast."[8]

"The first European believed to have touched North America was the Viking, Leif Erikson, who came by boat from Greenland to Nova Scotia—500 years before Columbus' 1492 voyage to the West Indies. There is no evidence that Columbus ever set foot on the North American mainland."[9]

"Other groups came by boats and ships to Florida from Europe and stopovers at the islands of the West Indies. "Florida was claimed for France in

1562 by their explorer, Jean Ribault and a group of Huguenots which were Protestants following the teachings of John Calvin. They established a colony at Fort Caroline near present-day Jacksonville. [That was twenty-three years before the Lost Colony in Virginia, forty-five years before the first perma-

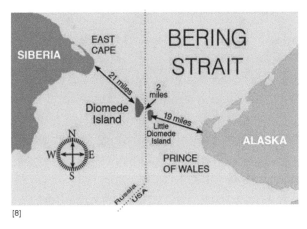

[8]

nent settlement in Jamestown, and fifty-eight years before the Pilgrims landing in Plymouth, Massachusetts.]

Even earlier, Ponce de Leon preceded Ribault in 1513 by landing on Florida's East Coast. He claimed the territory for Spain, but his subsequent attempt to establish a colony was abandoned due to hostile resistance from the native Calusa Indians.

The oldest continuously occupied settlement of European origin in the United States is St. Augustine, Florida, founded by another Spanish explorer, Don Pedro Menendez de Aviles, in 1565."[10]

The origin of the first pioneer remains in dispute, but uncontested is that North America is a continent founded by immigrants from far, wide, and parts unknown. The strongest evidence leans in favor of those of Asian descent who have been here thousands of years and who dispersed to every region from coast to coast and border to border—Native Americans.

All the immigrants left their homelands for compelling reasons. The forefathers of William Driver, migrated to the Massachusetts Bay area around 1620. They were Pilgrims and Puritans from Europe, with exceptionally strong religious beliefs, tendencies, and heritages—true pioneers that mixed and mingled with the descendants of other like-minded pioneers as well as those from different persuasions.

His ancestors—their roots, and their frames of reference—naturally influenced his stories, and their lives were models for his life as he became "a part of all that he met."[11] They came from England via Holland during an era of major religious upheaval and revolt against the Roman Catholic Church. Martin Luther had led the Protestant Reformation in 1517, creating the background and environment into which William Driver was born.

Much has been written about my great-great-grandfather throughout the decades: however, none had the resources that are in my possession nor had spoken with the last credible witness who knew him personally. This person was his grandchild and my Grandmother, Georgie Driver Reece Wade, who was thirteen before Captain Driver died.

Consequently, she is the primary source and basis for my frame of reference. It began in the early decades of my life while sitting by an open fireplace on Greenfield Avenue in the Inglewood community of Nashville, Tennessee. I returned to that favored place many times over the years to listen to Grandmother Georgie tell enchanting stories about our famous sea captain grandparent. It was a peaceful and trance-inducing setting looking into the burning embers flickering, popping, sparking, and spewing. Perhaps I also had carefree visions of being on board his ship sailing away to parts unknown. Little did I realize that one day, it all would come around again, and I, indeed, would take to the sea as the captain of my own boat.

This is more than a twice-told tale. I hope to tell it with greater accuracy, clarity, fullness, and appeal than ever mentioned before.—*Jack Benz 2019*

"There are no foreign lands. It is the traveler only who is foreign."[12]

—Robert Louis Stevenson

The Forefathers of William Driver

"England's Henry VIII broke ties with Catholicism in the 1530s after the Pope would not allow him to annul his marriage to Catherine of Aragon. His rationale was that she had failed to produce a male heir. Afterwards, he declared himself head of a new national church called the 'Church of England'—later 'Anglican.' Although he and his daughter, Queen Elizabeth I, changed some things that made the Church of England different from the Roman Catholic Church, some felt that the new Church retained too many practices of the Roman Church. They wanted a return to a simpler faith

The landing of the Pilgrims by Andrews[9]

and less structured forms of worship the way the early Christians had practiced. Because they desired to purify the church, they came to be known as 'Puritans.' Others wanted to separate completely into distinct unaffiliated and self-governing congregations; they became known as 'Separatists.'

"Technically, it was illegal to be part of any congregation other than the Church of England. Those who refused to follow its teachings were harassed, fined, or jailed. Consequently, they fled to the Dutch Netherlands in 1608, where they could practice their own religion without fear of persecution.

"After about a decade, the congregation worried that another war might break out between the Dutch and Spanish, so they decided to move again—this time to a farming village in the northern part of Virginia. They bargained for their transportation on the *Mayflower,* and 102 of them arrived in New England on November 11, 1620, after a voyage of 66 days. Although the Pilgrims initially had intended to settle near the Hudson River in New York, dangerous shoals and unfavorable winds forced the ship to seek shelter at Cape Cod. Because it was so late in the year and traveling around the Cape was proving to be difficult, the passengers decided not to sail farther but to remain in New England, primarily at St. Ann and Plymouth.

"A party of the ablest men explored the area to find a suitable place to settle. After several weeks, they arrived at what appeared to be an abandoned Wampanoag community. A plentiful water supply, good harbor, cleared fields, and located on a hill made the area an ideal place for settlement.

"The colonists began building their village while still living on the ship. Many fell ill, likely from scurvy and pneumonia caused by the cold and wet environment. Although they were not starving, their sea-salt diet had weakened their bodies on the long journey. During that first winter, as many as three people died each day. Only 52 survived the first year in Plymouth. The *Mayflower* returned to England on April 5, 1621, but with only half of her crew.

"It was here in Cape Cod Bay that most of the adult men on the ship signed the document known as the Mayflower Compact. It laid the foundation for the community's government. The Separatist church congregation that established the Plymouth Colony included the young William Bradford—a signatory to the Mayflower Compact and future Governor of the Plymouth Colony."[13]

The Native Americans

At the time of early European colonization attempts, there were over 30,000 Algonquin language Native Americans in Massachusetts. Archaeological evidence shows that Native Americans have lived there and on the adjacent islands for at least 8,000 years.

Although the colonists occasionally caught glimpses of the native people, it was not until four months after their arrival that they met and communicated with them. They established an amicable relationship and, in March 1621, they made a treaty of mutual protection with the Wampanoag Chief, Massasoit.

Myles Standish was hired by the Pilgrims as military adviser for the Plymouth Colony. He accompanied them on the *Mayflower* journey and played a leading role in the administration and defense of the Plymouth Colony from its inception. On February 17, 1621, the Plymouth Colony militia elected him as its first commander and continued to re-elect him to that position for the remainder of his life.

Massasoit and Gov. John Carver smoking a peace pipe being observed by Myles Standish [10]

The Treaty had five points: "Neither party would harm the other. If anything was stolen, it would be returned, and the offending person returned to his own people for punishment. Both sides agreed to leave their weapons behind when meeting, and the two groups would serve as allies in times of war."

"The natives instructed the colonists in growing corn, and in the fall of 1621, they united to celebrate their first harvest. Massasoit and 90 of his men joined the English for a feast which was to become recognized as the first Thanksgiving."[14]

"Unfortunately, the Europeans would bring with them diseases against which the Native Americans had no immunity, and which resulted in large deadly epidemics. The native population continued to suffer from disease and skirmishes throughout the remainder of the 17th century. Nearly ninety percent of the native population died during that period.

"Several 'praying towns' had been established within Massachusetts Bay where the natives tolerated living among their European neighbors. As colonial settlements expanded, many Native Americans were displaced to the Indian praying villages and towns.

"War ultimately erupted in 1675 between the colonists and the Native Americans led by King Philip, a title the colonists gave to Chief Metacom. The colonists overwhelmed them, and the survivors were interned to the east coast islands to face dire circumstances and deteriorating relations. Some were sold into slavery, or indentured into English households as servants.

"The voyages from England became more frequent, and the Pilgrim population grew to nearly 14,000 by 1640. The first baby was born—a boy—to parents William and Susannah White. They named him Peregrine—a word meaning traveling from far away, the same definition as the word pilgrim. Peregrine followed Virginia Dare of the Roanoke Lost Colony to be the second child of English parents born in the New World."[15]

In summary, this incredible experiment and resulting "Great Migration" was placed in perspective by John Winthrop, a Puritan layman, lawyer, and future Governor of the Massachusetts Bay Colony.

Winthrop delivered a sermon about his vision to the passengers aboard the ship *Arbella* in 1630. It was one of eleven ships carrying over a thousand Puritans to Massachusetts and was the largest early venture attempted in the English New World. He asserted that they were called to be an example for the rest of the world in rightful living: "We shall be as a city upon a hill; the eyes of all people are upon us." His obvious analogy was taken from the Biblical reference in Mathew 5:14 and is considered to be one of the most repeated utterances of all times. The passengers embraced his powerful words and became determined to be a beacon for the rest of Europe—"A Model of Christian Charity."

Salem, Massachusetts
A Perspective on its History

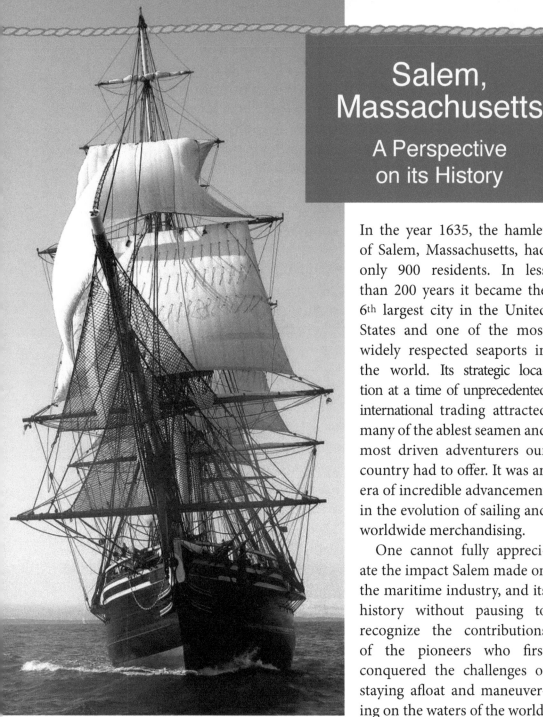

The *Friendship* Underway [11]

In the year 1635, the hamlet of Salem, Massachusetts, had only 900 residents. In less than 200 years it became the 6th largest city in the United States and one of the most widely respected seaports in the world. Its strategic location at a time of unprecedented international trading attracted many of the ablest seamen and most driven adventurers our country had to offer. It was an era of incredible advancement in the evolution of sailing and worldwide merchandising.

One cannot fully appreciate the impact Salem made on the maritime industry, and its history without pausing to recognize the contributions of the pioneers who first conquered the challenges of staying afloat and maneuvering on the waters of the world.

In some respects, Salem paralleled Sagres, a seacoast town in Portugal that two-hundred years earlier was considered to be an advanced center of nautical knowledge and idea exchange. It is where Prince Henry, the Navigator, founded a school which was more of an arena for the study of astronomy, geography, mapmaking, mathematics, navigation, shipbuilding, sailing techniques, and trading skills. Among the more notables associated with Prince Henry were explorers Christopher Columbus, Vasco da Gama, Bartolomeu Dias, and Ferdinand Magellan. The latter erroneously is credited to be the first and youngest person to circumnavigate the world. However, because Magellan died before completing his trip, many believed Salem's William Driver ultimately will be confirred that distinction. Regardless, Salem flourished in the wake of many predecessors, and its prominence on behalf of the seafaring is enormous and will endure forever.

A look into the past at the evolution of boats is appropriate. It is a common observation that most youth learn about principles of buoyancy at their swimming holes. They find large limbs and logs floating on the water functioning similar to crude boats. The swimmers holding on to them find that neither sinks to the bottom. Their teachers explained that the reason is in accord with Archimedes' principle, i.e., if the weight of water displaced is greater than the object replacing it, the object will float. His principle has proven to be accurate and infallible even if the object weighs tons. It is the fundamental premise of all shipbuilding.

"Drawings of ships along with crude building tools have been found dating back to the Neanderthals 100,000 years ago."[16] Actual canoes up to 16,000 years old are preserved in museums in Korea, the Netherlands, and Nigeria. Paintings on pottery found in ancient Mesopotamia, dating back to 5,500 B.C., depict large boats afloat on the Nile River. The vessel Khufu is 4,500 years old and is housed in Giza, Egypt. In 700 B.C., Homer wrote about shipbuilding and sailing in his epic Odyssey.

There are over 40 Biblical references to boats with Noah's Ark in Genesis 6:9 being the most prominent. Later in Exodus 2:3, Moses' mother hid him in a floating basket she fashioned from papyrus coated with tar and pitch—a boat.

The forefathers who helped settle Salem arrived aboard ships as the first Pilgrims and Puritans in the New World. It is natural that their heritage, value system, and mindset became the frame of reference that molded William Driver's entire life and thought. Merchant shipping was the main course of business; everything else was saturated with and by religion.

"Salem is a Hebrew word meaning peace. It was founded in 1626 by Roger Conant and immigrants from nearby Cape Ann. Two years later, they were joined by another group led by John Endecott, who became the longest-serving governor of the Massachusetts Bay Colony.

"During the 1776 War of Independence and the War of 1812, Salem was a sanctuary for privateers. These were privately owned, manned, and armed ships hired by the government to fight enemy ships. This practice represents one of history's earliest examples of outsourcing services to a private contractor.

"Salem ship captains took their vessels to distant ports and earned great wealth for themselves, their owners, and their city. The original *Friendship* is typical of the fleet of merchant ships ported at Salem Harbor in the early 1800s. It made 15 voyages to exotic ports, including China, South America, Germany, the Mediterranean, and Russia. It was not uncommon to profit enough on a single journey to cover the entire cost of the vessel. She was captured during the War of 1812, and her fate continues to be unknown.

"During its maritime heyday, Salem helped build the nation's economy focusing it's trading on Atlantic Cod, brandy, cotton, fruit, iron, ginger, rum, salt, silk, sugar, tea, and wine." Many spices literally were worth their weight in gold.

"Although Salem is known worldwide as a major seaport, a more infamous association, unfortunately, is the more common recollection. A witchcraft hysteria gripped the area in the closing years of the 17th century. A local physician diagnosed several teenage girls as bewitched. Tituba, one of the accused, actually confessed that she and other witches were working for the Devil. This admission resulted in several being imprisoned, ten being hanged, and one being crushed to death."[17]

Perhaps the context of the times and circumstances explains how such a horrid event occurred. The community was in a deep state of anxiety because King Charles II had revoked their charter due to religious revolt against the Church of England. As well, the town was born and nurtured by strong fundamental religious influences, beliefs, and practices. The Devil was real and believed to be-

Witchcraft at Salem Village [12]

coming entrenched in the colonies; so a witch's confession was all that was needed for the hysteria to erupt. Many were placed in prison, but fortunately, sensible people resolved the matter and released the imprisoned, thus ending witchcraft trials in America forever.

The preceding happened over a century before William Driver's birth. However, he never could escape endless disparaging conversations prompted as a result of Salem's infamous legacy.

The early seafarers had to learn and master certain basics to survive in the domain of the oceans and seas. They learned to cooperate with gravitational forces and to harness the currents, temperature, and wind. In the absence of land-marks, they studied makeshift charts, crafted and improved by experience from generation to

The Early Mariners

generation. They relied on crude instruments, celestial bodies, geometry, and mathematics to guide their ships to known destinations and to discover new ones. William Driver was such a discoverer, and his captivating adventures are portrayed throughout this book.

The Light of Navigation, Dutch Sailing Handbook 1608 [13]

CHAPTER 1

His Birth and Setting

...out of the lands from the east, and from the west, from the north, and from the south...he brought them out of darkness and the shadow of death...they that go down to the sea in ships, that do business in great waters[18] – Psalm 107

Salem Harbor

My earliest memories are from my first breaths and sounds. The ocean breezes brought that distinctive briny scent to my crib there by the Bay in Salem, Massachusetts.

It is that unmistakable sulfur-like smell familiar to all who dwell near or sail upon the salty waters of the world. My teacher, Professor Isaac Hacker, believed the aroma was caused by a chemical reaction involving decaying algae. Regardless of the source, I find a mere whiff to be intoxicating and medicinal. It clears the nostrils and mind like no other potion. More so, it draws one like a magnet to the great waters.

Another of my senses awakened during my first days living by the sea in the Massachusetts Bay Colony. The sounds of the surf caused by the rising and falling of the tides lulled me to sleep and continue to echo in my ears to this very day. The roaring and crashing of the waves subside after a while but routinely cycle again and again in an endless procession.

There is even more in Salem to soothe the mind, body, and spirit. The flags and sails on the ships coming and going in the harbor flap and pop in the brisk winds like a snare drum cadence. Together, they provide a rhythmic cure for sleeplessness. The joyful voices from the hands-on-deck preparing to dock or embark, join with the laughter, cheers, and tears onshore to announce that all is well.

Indeed, those first recollections from my kingdom by the sea remind me of how fortunate I am to have been born in Salem, Massachusetts. For me to better understand who I am, where I am going, and why, I often reflect on how the circumstances of my birth and environment profoundly influences my life and thought.

The United States of America was in its infancy in 1803—only 27 years old. However, the fighting during the Revolutionary War did not

Surrender of Lord Cornwallis [15]

23

end with our Declaration of Independence. It officially occurred on October 19, 1781, when General George Washington's Continental Army, aided by French troops, defeated General Cornwallis and the British Army at Yorktown.

It is of further significance to note that thirteen years earlier in 1763, France ceded all claims of North American Territory east of the Mississippi River to Great Britain as part of the settlement ending the French and Indian War. Consequently, when the United States was granted its independence, this additional territory was included in the agreement. Fortunately, it enabled the colonists to explore and settle far beyond its original boundary created by the Appalachian Mountains.

Three additional states—Vermont, Kentucky, and Tennessee—already had joined the original thirteen by the time of my birth. Further, Thomas Jefferson's "Louisiana Purchase" from France included all the land from the Mississippi River west to the Rocky Mountains and from the Gulf of Mexico north to the Canadian border. This purchase acquired 827,000 square miles of new territory and ultimately created 15 additional states at a cost less than four cents an acre.[19]

The magnitude of these world-changing events and their impact on my young life are staggering. Thirteen struggling immigrant-driven colonies from 1776 to 1812 defeated the mighty British Army—reputed to be the best and most professional in the world; they defeated Britain's Royal Navy, the undisputed ruler of the high seas; they achieved total independence as a nation and doubled the size of its territory—all in a span of 36 years.

Naturally, I was unaware of most of the foregoing. I was a mere four-year-old pre-school child doing things my Puritan-minded parents led me to do. Excessive playing by the youth was viewed as an abomination, although I did have a few hand-made wooden toys and was provided occasional free time to play with other children on the endless sandy shoreline. My preferred pasttime was watching the nonstop sailing-related activity in the nearby harbor. I was trusted to sit on a large shaded rock near the wharf, sometimes for hours, while my parents were at work or conducting business. As long as I did not wander off, my rock was a refuge and relief for me to escape the rather boring and rigorous routine I had grown to accept as my lot in life.

After all, what kind of young child could enjoy sitting still all dressed up in itchy wool pants, being polite to the elderly, and performing unexciting chores? It was further punishment hearing about family history, debates over religious doctrine, attending funerals and church services. The most distressing of all, listening to hours of frightening sermons from shouting ministers. These horrible activities seemed to be the major preoccupation for all the families in that day in the Massachusetts Bay Colony.

New France, 1750

Control of the Territories
- France
- Territories ceded by France to Great Britain by the Treaty of Utrecht, 1713
- Great Britain
- Spain

New France [16]

As I grew older, I came to understand why most of these practices were so important, and perhaps explain why I embraced many of them with my own children. The driving force and controlling influence woven throughout my early nurturing continued to be religion."[20]

"My parents and grand-parents' spiritual roots were influenced by the Pilgrims and Separatists of the Plymouth Colony and the Puritans in the Massachusetts Bay Colony. As previously addressed, both groups had evolved in the old world from Roman Catholicism, through the Protestant Reformation to the Church of England (Anglican). The latter group wanted to stay connected with the Church of England but to purify it; the Pilgrims wanted to separate completely.

"After that, they reformed further and became 'Congregationalists,' meaning that the memberships ran their own church affairs independent from the authority of the Roman Catholic Pope and the Anglican Archbishop of Canterbury. My family continued to engage in reform movements and ultimately most became Universalists—those who believe that all human beings eventually will be granted eternal life."[21]

My religious evolution continued when two of my brothers became Baptist Preachers. Both tried to steer me away from the Universalist Doctrine but to no avail. I was much too young and unimpressed with religion in general. In fact, I had developed some strong negative opinions about God, Christians, church, and all preachers. Their notion that I should become a minister did not suit or fit my interests at all. The long hours I spent in church enduring agonizing sermons and foreboding preachers erased that choice permanently.

First Church in Salem Village [17]

CHAPTER 2

School Days - My First Jobs

First Boston Latin School [18]

I feel fortunate to have been born close to Boston only 18 miles from Salem. America's first and oldest continuing public school, The Boston Latin School, was founded there in 1635.[22]

The school thrived, and just twenty-seven years before my birth, five of its former students signed the Declaration of Independence. They were Samuel Adams, Benjamin Franklin, John Hancock, William Hooper, and Robert Treat Paine. Only Benjamin Franklin of the five failed to graduate. Looking forward, I find it strange that only the dropout has a statue honoring him gracing the front of the school.

In 1645, the Massachusetts Bay Colony made public education a requirement for all students. Salem was one of the first towns in America to fully implement the law. The strong emphasis Puritans place on children learning to read and write unmistakably is so they can study the Bible and civic law. There is no doubt that Boston's lead in the movement is responsible for its reform and rapid spread and throughout the colonies. I continue to find the influence of religion gaining momentum and moving beyond the pulpit to the classrooms and school books.

In *Adam's* Fall
We Sinned all.

Thy Life to Mend
This *Book* Attend.

The *Cat* doth play
And after lay.

A *Dog* will bite
A Thief at night.

An *Eagles* flight
Is out of sight.

The Idle *Fool*
Is whipt at School.

Reading Lesson [19]

An interesting coincidence occurred when I retired from the sea and moved to Nashville, Tennessee. Hume-Fogg High School was located on Broad Street, a block away from my church home, Christ Episcopal.

"In 1852, before the school was built, Nashville officials, considering the establishment of its first public school system, sent Professor Alfred Hume up East to observe several exemplary school districts to determine the best model to follow. The Massachusetts Bay area schools prevailed in convincing Professor Hume that public education for all was essential if the nation was to survive; and that the ways and means—curriculum, decorum, philosophy, and methodology practiced in the Boston area schools were the most effective he had seen. He returned to Nashville and implemented them accordingly."[23]

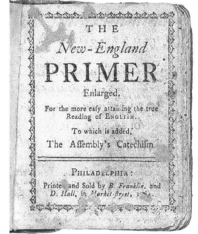

New England Primer [20]

"Fortunately, this same thinking and same notions were embraced by my teacher, Professor Isaac Hacker in Salem. He taught me broadly in the classics and prepared me fundamentally. According to him, I excelled. I was under his command and influence from age five through twelve until I became an apprentice at Mr. Goodhue's Blacksmith Shop.

"In a sense, my school experience seemed very much like a Puritan church service. There were similarities in the tone and manner of the membership as well as the dispositions of the teacher and preacher. I do not recall ever seeing a smile from either. Discipline was strict—bordering on harsh—and foolishness was not tolerated. Professor Hacker was a taskmaster sparing not the rod for misbehavior, tardiness, or lack of studiousness."[24]

"My parents, like most in the colonies, taught me basic reading and writing at home before I attended school. Regardless of preparedness, the course of studies at The Hacker School was the same for all students, i.e., Reading, Writing, Arithmetic, Grammar, Latin, Logic, Penmanship, Rhetoric, Geometry, and Astronomy. As well, decorum, good manners, and Christian principles were interwoven throughout the curriculum. It was a rigorous challenge, and though I did not realize it at the time, Hacker's greatest gift to me was not what I learned, but that he taught me how to learn; I became a continuing learner for my entire life.

"The eighth grade was considered to be the terminal level of education for most students, which happened to be boys. Girls conspicuously were absent until the Women's Suffrage Movement began in the mid-1800s.

"My forefathers arrived from Europe strongly indoctrinated in Puritan principles and practices. I do not recall the Puritan work ethic being discussed or defined in my early childhood, but I quickly came to understand its meaning. Like everything else in my existence, religion was the resource used to guide all words, thoughts, and deeds. The Bible was the rulebook, and from the pulpit it was explained—quite harshly I might add. The interpretations of Martin Luther and John Calvin were deeply rooted and embraced by my parents, Stephen and Ruth Driver, much the same as they were more than two centuries before my birth. The foundational beliefs, particularly for children, emphasized the high priority that God's Word places on the importance of hard work.

"Pilgrim children were taught life survival skills, running errands, gathering wood, carrying water, and herding chickens. The older boys helped their fathers prepare the fields for planting, sowing, weeding, and harvesting. The mothers taught the girls

skills such as gardening, cooking and preserving food, tending to the younger children, and sewing."[25]

My parents were strongly attuned to Puritan work ethic principles and teachings. They were very strict disciplinarians in all aspects of my training and performance of my chores. Mother particularly emphasized that I do my very best in every endeavor. Occasionally, they would bring up the subject of my "calling"—my future profession.

I wanted to become a sea captain, but my parents were fearful, for a good reason—hardships and danger. My ship Captain Grandfather, Michael Driver, had experienced several bad incidents while sailing his schooner in the late 1700s. He had escaped several takeover attempts by pirates, but another significant circumstance and event took his life. Our country was at war with France—second only to Great Britain as the largest Colonial empire in the world.[26]

Sailor Boy [21]

US Constellation and French frigate *L'insurgente* during the Quasi-War [22]

It was known as the Quasi-War of 1798—fought entirely at sea and only involved the United States and France. It lasted for two years but proved to be an unfortunate time frame for Michael Driver. A French boat riddled the entire broadside of his schooner, causing him to retaliate with a canon shot through the port quarter of their boat. Apparently overwhelmed, his schooner was taken into custody, and during the questioning of my feisty grandfather, his poor choice of words resulted in a very bad outcome. When the French officer asked him about the damage from his cannonball, he responded, "I wish it had been your Emperor." The intended insult got him thrown into the infamous French prison at Pointe-a-Pitre, Guadalupe, where he remained until he died.

Therefore, it is understandable why my parents tried to steer me away from the sea. My father's first attempt was an unappealing offer to teach me his cobbler trade. Later, perhaps to please them, I reluctantly agreed to become an apprentice at Mr. Abner Goodhue's Blacksmith Shop. My job was to pump the bellows to keep the fire red hot. Since I was short in stature, I had to stand on a candle box to reach the bellows pole.

To make a boring job worse, Mr. Goodhue added another task. I had to milk his old cow, Brundel, twice a day and deliver the milk to a route of regular customers. Many were notable sea captains who owned some of the more beautiful homes in Salem. I was impressed by their standard of living and could imagine that as a possibility for me someday.

I did not mind the hard work, but I had to endure a lot of bullying from the

Bellows[23]

neighborhood boys. I had a real problem frequently being called "Smitty Boy," referring to my blacksmith job. That taunt usually started a rock fight. More importantly, I did not see that being a blacksmith was an exciting and adventurous job. It is honorable and important for sure but not a good fit for me.

Child Milking Cow[24]

One particularly unpleasant Sunday morning—my day off from work—I was staying with Mrs. Goodhue, who dressed me in a muslin shirt with scratchy ruffles and sent me off to Sunday School. I rebelled by cutting off the ruffles and headed straight to the docks. As I approached, I was thrilled to see the big ship, *China*, loading for another glorious trip to sea. The harbor was buzzing with excitement and enthusiasm as the sailors prepared for departure. A stiff breeze from the east cleared my head and boosted my confidence as I marched up the steps of the Custom House, hoping to sign on as a Cabin Boy. Somehow I knew, if only I could get on that ship, I would find the path to fulfill my dreams.

CHAPTER 3
The Cabin Boy Goes to Sea

Quay at Honfleur [25]

I entered the custom house somewhat nervous, but reasonably confident in how I was going to handle myself in the interview for the Cabin Boy job. The sailors hanging around the wharf waiting to ship-out had prepared me for what to expect. As well, Professor Hacker, along with my parents, had taught me about proper etiquette with adults and especially when meeting new people—eye contact, a firm handshake, and an energetic voice. I figured that there were other boys wanting the same job, so I needed to present myself in a more mature and confident manner to improve my chances of being chosen.

Some sailors were standing in a line behind a desk waiting to speak to a well-dressed gentleman, obviously the person doing the hiring. A sign identified him as Mr. Tucker. I could overhear the conversations as I moved closer to my turn. Some were being hired, and others turned away. It seemed that the ones with outgoing personalities were having better success than those who were withdrawn and timid; I was prepared as well as I could be.

When my turn came, I smiled, stuck out my hand and asked, "Sir, is there still time to sign on to the *China* as the Cabin Boy?" Mr. Tucker answered "yes" but questioned my age since I was shorter than average for my years. I told him I was fourteen, had lots of work experience with good references, and could assure him if he would hire me, I would be the best Cabin Boy ever. Mr. Tucker seemed impressed but tested me further by revealing that the pay would be only $5.00 per month. That was because his experience with Cabin Boys had taught him that most do not even earn the cost of what they eat on their first voyage. That only made me determined to show him that there is one boy in the world who indeed can earn much more than the pay for what he eats. I told Mr. Tucker that he was looking at him, so he hired me immediately.

I was overjoyed that soon I would be going to sea on the famous *China*, one of the swiftest ships on the ocean. It was owned by Mr. Tucker and his associate, Mr. Peabody. Mr. Hiram Putnam was to be the shipmaster—the person responsible for navigation. Our destination was Leghorn, Italy, and our voyage was scheduled to last for 18 months. After hearing about my good news, my parents' fear seemed to lessen because of the outstanding reputation of the *China* and its crew.

I knew a few things about Leghorn that I learned from returning sailors and from my grandfather, Captain Michael Driver, through my father. Furthermore, I went to Professor Hacker's School and found several books including an Atlas that gave me a wealth of information about where I was going—my very first journey away from my home and parents. "I learned lots of interesting facts, including one about "Leghorn"—an English word—but correctly translated as Livorno. It is a coastal city in a large region known as Tuscany. The leaning tower of Pisa is close by as is Florence, the capital which is considered

I learned lots of interesting facts including one about **"Leghorn"** —an English word —but correctly translated as Livorno.

Leghorn

Livorno, Italy [26]

to be the birthplace of the Renaissance. Nearby Elba is the isle where Napoleon was exiled and later escaped. Even Professor Hacker would be excited about this opportunity to live and work amid such remarkable history."[27]

Clearly, I was headed to a cultural center of the world and a retreat paradise, but that was not the reason for such a massive undertaking. I learned that crossing the entire Atlantic Ocean and passing through the Strait of Gibraltar to the shores of Western Italy presented a far more compelling reason. It was because Livorno was one of the most favorable trading centers in the world due to its "Free Trade Policy."[28] It had the reputation of being unusually trading friendly to foreign merchants in terms of the costs and limitations on conducting business.

Ships from the east and the west converged there loaded with their wares and specialized commodities. They sold or traded one product for another that would be in demand on the return trip. Spices, including cloves, cinnamon, pepper, and nutmeg were highly prized. Also in demand were precious metals, jewels, and other commodities such as coffee, tea, sugar, guns, textiles, tools, china, perfumes, silk, alcohol, furs, and tobacco—and the list seemed endless in this golden age of bartering. It was an extremely profitable enterprise, especially for the ship owners and captains. For example, I was told, "when the schooner *Rajah* sailed into Salem Harbor in 1797, her hold was packed with wild Sumatran pepper—used as a meat preservative and valued at about $125,000—seven times the cost of the vessel and her contents when she had left Salem 18 months earlier."[29]

When I first learned that this trip would last 18 months, I did a little figuring. It would take us about four months going and four months returning, leaving ten months for our trading. I was puzzled why that much time was needed until I learned that the Captain often bartered during side trips for extra money. Nevertheless, those eighteeen months were among the most eventful and educational in my entire life. The work was hard and the days long. As the youngest member of the crew, I was given the chores no one else wanted. I was at the beck and call not only of the Captain but every sailor on the ship. Cabin Boys were often disadvantaged and discouraged from ever taking another voyage. If they were truing to discourage me, their efforts were in vain. I now knew without a doubt that the sea would be my life's workplace.

We made it back to Salem safely, and I reported to Mr. Tucker as requested. With a hearty smile and handshake, Mr. Tucker handed me $71.80 for my father and kept $32.00 for the clothes he furnished for the voyage. I understood that this was the customary practice when minors were involved. Then Mr. Tucker counted out twenty-eight Spanish silver dollars and handed them to me saying, "this is for you, my boy, as a reward for being the first boy ever known to have earned on his first voyage enough to pay for what he eats."[30]

With many thanks and smiles, I ran home as fast as possible to show my silver dollars to mother. She was as proud of them as I and encouraged me to continue always to do my best. I never forgot that moment or my commitment to do so until my last breath.

CHAPTER 4
Off to Calcutta India

Bathing in River Hoogly [27]

The Great Awakening for my forefathers mainly related to religious reformation and discernment. The awakening of students by Professor Hacker was quite different—far more than teaching fundamentals of the reading, writing, figuring, and speaking. He stressed the enlightenment of our minds beyond the classroom to the taking advantage of daily opportunities to engage with intelligent people, challenging conversations, new frontiers, and domains that broadened our understanding of the universe and its intricacies. He challenged us to be continuous learners; to seek wise counsel and wholesome friends that strengthened us; and to avoid companions that know little, behave poorly, and care less—those who by association tend to pull us down.

I never fully appreciated his insight and sensible advice until I "took to the sea." I rapidly learned the basics of seamanship from my on-the-job-training to and from Leghorn. More importantly, I discovered a vast reservoir of knowledge and skills available in the heads of experienced mariners, and in the people living in faraway places reflecting different cultures and practices.

The difficulty of navigating a wind-driven merchant ship thousands of miles and conducting complicated business transactions upon arrival is enormous. It is not a job that ordinary folks can fulfill. Likewise, they confirmed Professor Hacker's urging about choosing my companions; simply associating with exceptionally talented and gifted individuals is uplifting—a rising tide lifts all boats. Homer provided a parallel in his travels of Odysseus.

> "Many were the cities he saw, Many were the men whose minds he learned, And many were the woes he suffered on the sea."
>
> The Odyssey - Homer

Consequently, my orientation and experience gained from the voyage to and from Italy greatly enhanced my roles, responsibilities, and demand for my services. Accordingly, I am honored to recently have been offered and accept two assignments to and from Calcutta, India. I will be sailing on the *George* and serving under Captain Endicott, an accomplished Massachusetts Bay Colony mariner. I am preparing for this journey by reading widely and speaking broadly with those who are eye witnesses—those already having been to Calcutta.

I am eager and excited to visit my third continent, Asia, to sail on my second of the world's three largest oceans, the Indian, and to be exposed to the second and third of the world's three largest religions, i.e., Hinduism, and Islam—still as a youth. "What hath God wrought for me at such a young age!"[31]

Governmentally, I learned that India had been under foreign domination for over a thousand years and effectively under British rule for nearly 200 years.[1]

---

[1] A look forward revealed that it was 1947 when this great country and its millions of people finally became a sovereign nation. Britain's House of Commons passed the Indian Independence Act which divided India into two countries, India and Pakistan. At this time,2019, both are among the world's nine nuclear weapons states.

Fort William[28]

India's Connection with Joint-Stock Trading Companies

Joint-Stock Companies played a major role in the discovery and development of colonies and trading networks around the world during the seventeenth, eighteenth, and nineteenth centuries. These companies financed the equipment, materials, manpower, and ships by selling shares of ownership to investors seeking profits from the various ventures. The companies were sponsored to a degree by their native country's rulers seeking to expand their control and influence, primarily in trading, although in some instances colonization was their goal. Thereby, these various investor groups were granted "Charters" which protected them from competition from other countrymen, but not from other countries—they had a monopoly of sorts.

I have no desire whatsoever to participate in colonizing ventures. My interest is in trading commodities like my grandfather, Captain Michael Driver, with one major exception. That is a venture that in my future happened by chance and coincidence, yet, it altered my life and destiny.[1]

I was encouraged by Captain Endicott to research how joint-stock trading companies have impacted India and influenced its commerce. He requires all

---

[1] It is a "transporting venture" that in my future happened by chance and coincidence; yet, is destined to alter life and thought. It happened in Tahiti about which in time the reader will learn in great detail.

his sailors to be generally well-informed so we never will be disadvantaged on foreign soil. There is a small library on the *George* containing general knowledge books to help us in that regard. He even encouraged me to begin learning a basic level of the native language spoken in India, as well as those in other countries I might visit in the future. He emphasized that fluency in several languages would be very good for my career and might even save my life someday

"I was shocked to learn that over 20 languages are spoken throughout India, but Bengali, Hindi, and English are the most common. The Captain loaned me his personal language translation manual that teaches common expressions used with basic communications in several foreign languages. I chose to focus on Bengali after learning that it is considered to be the sweetest and most romantic language in the world. A couple of my shipmates could speak and understand it; indeed, it does have a beautiful rhythmic sound, and they promised to help me with it. Further, in my off-duty time, I continued strengthening my knowledge of the joint-stock trading companies I might encounter as competitors on this and future voyages."[32]

I found that all the early colonies in North America were established as a result of these venture capital companies, e.g., the Hudson Bay Company, the Virginia Company, and the Massachusetts Bay Company.

"Before 1600, India was ruled by Mughals, a Muslim dynasty extending back to Genghis Khan. However, the Spanish and Portuguese maintained an independent monopoly on the East Indian spice trade.

"Thereafter, the British East India Company in 1600 and the Dutch East India Company in 1602 were chartered and became formidable competitors. At first, the Dutch were the dominant force in international trade, but something unusual had occurred within the wording of the Charter granted to the English merchants by Queen Elizabeth I. The British East India Trading Company was given authority to colonize—to take over a country. It even had an army and ruled like a sovereign nation over all of India's affairs from distant London. This continued throughout my life and explains why the English language continues to be in wide use there today. The British continued to colonize territories worldwide throughout the 19th century, becoming the largest empire in history. Understandably, it became 'the country on which the sun never sets.'

"The East India Trading Company had a massive private army of one-quarter million soldiers—double that of the entire British Army and which oversaw half of the world's trade."[33]

Now I understand the insistence on me learning all about joint-stock trading companies as an important part of my overall dossier. My success in becoming the captain of my own ship will depend on my ability to convince the owners and investors of my knowledge and skill far beyond seamanship.

Dutch East India Trading Ship [29]

They must be confident that I fully understand the "Merchanting Trade." Captain Endicott gave me a check-off study list to help strengthen my qualifications. It included areas related to commodities and products controlled by the owners, countries specializing as suppliers of high demand items, world trading centers serving as intermediaries, economic laws and principles, intricacies of stocks and bonds, futures contracts, and derivatives. I must understand and be able to apply differential calculus used by captains and trading officers to predict the movement, direction, and rate of change

British East India Charter [30]

occurring in the markets. Mercy! I have come to conclude there is much more to this business than meets the eye; indeed, it is not leisurely sailing about. I eagerly accept the challenges and am responding accordingly.

It is a coincidence that some of my Pilgrim forefathers spent time in Holland as Separatists before their 1620 voyage to New England. The oldest stock exchange in the world was established there in 1602 by The Dutch East India Stock-Trading Company. The New York Stock Exchange did not begin until 1792 just eleven years prior to my birth. Fortunately, I am in on the ground floor—the early history of the learning curve. I hope that my experience, preparedness, and work ethic will negate any concerns about my chronological age and expedite my transition to the command of my own ship.

East India Military [31]

"There is a parting shot regarding the British East India Company. Just thirty years prior to my birth, it became indirectly involved in starting the Revolutionary War in North America's thirteen colonies. The Tea Act was passed by the British Parliament in 1773, granting it a monopoly on all tea sales in the Colonies. The colonists revolted and dressed as Indians, dumped 92,000 pounds of tea from the company's three ships into the Boston Harbor."[34]

The British retaliated by shutting down the harbor and passing laws known as "the Intolerable Acts."[35] Therein were demands of payment for the tea, quartering rights for their soldiers throughout the colonies, limiting town hall meetings to once a year, and requiring trials involving loyalists to be conducted on British soil. George Washington termed the Intolerable Act "the Murder

Boston Tea Party [32]

Act" because it would allow British Officers to harass citizens and escape justice.[1]

Land Ho

Hark! I hear a familiar shout of jubilation coming from the ship's watch. Land ho!... Land ho!... Land ho!..." Eureka, we soon will enter the harbor at Calcutta, India—known as the "Jewel in the Crown of British India." There is a mood of excitement and anticipation on board as we notice the usual appearance of the dockside workers busily preparing for our arrival—strangers from afar, we were to them—and a new and much different world for many of us.

I was not informed as to the cargo in the hold of the *George*, nor was I and most of the crew allowed to enter for strategic reasons. An untimely leak about our goods could negatively impact prices and trading strategies. I learned that the practice was a customary safeguard.

Typically, the contents depended on the destination and product we likely would carry back on our return trip. The bartering system commonly involved direct selling, trading, auctioning of products stored in the Calcutta warehouses, and outright purchases. A typical voyage might involve multiples of the preceding transactions without ever leaving the dock.[35]

Supply, demand, seasonal, and weather-related factors are the primary influences on value and ultimate price. The business is somewhat of a gamble, but well-informed and creative captains who are shrewd and prudent traders command enormous fees for their services. This can include percentages of the profits, lump sums, or combinations of both—in either case—a handsome reward.

I learned the most common commodities were cashews, china, cotton, coffee, hides, indigo and other dyes, gold, jute, leather, oilseeds, opium, rice, rubber, silver, spices, sugar, tea, textiles, tobacco, wheat, whiskey, and wool. I guessed that we had tobacco on board because of seasonal factors back home, but the captain kept that secretive among his chief officers until he assessed the pros and cons of the marketplace.

Every ship captain I heard about or had known became very wealthy early in life. They built fabulous homes with incredible views of the sea, and retired

---

[1] The Tea Party marked the beginning of the end of the British East India Company. It was abolished in 1874.

Return Visit of the Viceroy [33]

in luxury as young men. I had some trouble relating to that because I worked very hard on my first voyage for only five dollars a month. I came to better understand this after a valuable economics lesson from an old salt on board the *George.*

"He was a spice expert and showed me a small bag of saffron threads which are the stigmas and styles harvested from saffron crocus flowers. They are used for incredible seasoning, coloring, and medicinal purposes—mainly as an aphrodisiac—a favorite of Cleopatra. It blooms for only one week a year, and it takes 70,000 blossoms to yield a single pound. Because it is so labor-intensive to harvest and land-demanding, an ounce of saffron generally is more valuable than an ounce of silver. There have been periods in history and markets where the demand was greater than the supply, driving its cost per ounce higher than that of gold."[36]

That lesson convinced me if I succeeded in becoming a captain, I would have a good chance of becoming financially independent early in life. The risk-reward of international trading became very appealing to me, living on very limited resources. The love of money is not the reason for my motivation, but the desire to do the best I can energizes me to continue preparing intensely for whenever and wherever opportunity knocks.

The two trips to India seemed like taking courses in advanced education.

43

[35]

Saffron[34]

44

The first was business and professional; the second was cultural and historical. Professor Hacker, wherever you are, I listened to you, and indeed I am a continuing learner.

The months involved in crossing the Atlantic Ocean gave me daily on-the-job training in every aspect of seamanship. This included all kinds of rigging, securing, and repairing of every inch of the ship along with its equipment; learning how to make endless adjustments to the lines, sails, and booms; learning numerous nautical terms that require immediate potentially life-saving responses; gaining navigational knowledge using calculus, general math, geography, charts, dead-reckoning techniques, determining latitudes and longitudes; learning celestial and instrument navigation; becoming adept using an astrolabe, compass, chronometer, hourglass, quadrant, octant and sextant, learning much about tides, currents, wind and the ship's pilot wheel and rudder. Captain Endicott took me under his wing and nurtured me constantly. I looked to him as a father-figure while many on board were clearly intimidated just to be in his presence.

My second advanced study involved spending all my spare time learning about the culture, economics, geography, government, history, language, religion, and general way of life of the people of India. The constant echo of voices in my head came from my parents, Professor Hacker, Shipmaster Putnam on the *China*, and now Captain Endicott. "Prepare! Prepare! Prepare yourself broadly and widely in every domain and for every opportunity and encounter." Already, I have seen the fruits of their wisdom. Preparation and practice were giving me uncanny confidence and maturity well beyond my years.

Between the trading days and auctions, I was able to travel to many interesting sites of cultural and historical significance and to observe many unusual customs and curiosities. My very first walk through the market area of Calcutta was an eye-opener.

Houses of worship in India[36]

Spice Market

I was able to practice some basic expressions in the Bengali tongue, learned because of Captain Endicott's encouragement. It brought me instant acceptance by those surprised that I was bilingual as were most of them. Most everyone spoke Bengali but chose English as their native tongue. I concluded that was the case because their country was under British rule for centuries.

This day of sightseeing also took me to the sites of incredible worship houses of the three major religions practiced in Calcutta, e.g., Hinduism, Islam, and Christianity. They were known as temples, mosques, and cathedrals. It was my first exposure—of many to come—to a setting where my religious faith was in the minority; and where most of the people held spiritual beliefs, doctrines, and faiths uniquely different from mine. They were as fervent as I and as convinced that their paths correctly guided them in righteous living and to eternal rewards. This experience was not without serious risk!

Captain Endicott, along with many of the ship's crew, had schooled me about customs and practices that were taboo in India and in other Hindu and Muslim cultures. He advised that if I were to become a successful international trader, I must always present myself in a favorable manner and speech to other traders and to avoid being disrespectful. He forewarned that without knowledge and understanding of forbidden practices, I innocently could make a mistake that would destroy an immediate trade or worse—hurt my future advancement. He gave some prime examples that I embraced, which, in a way would save my life in an encounter with cannibals.

The Captain chose a simple illustration related to shoes and feet. In the foregoing cultures, shoes are considered to be dirty due to being in constant contact with the ground; because centuries of tradition have exaggerated their offensiveness, they always should be removed when entering houses of worship or homes. Sitting with crossed legs where the soles of the feet are exposed is considered to be highly disrespectful. Throwing a shoe at another is the supreme insult and invitation to fight.

Touching and asking personal questions, even if very discreet, are considered to be rude and an unwarranted invasion of one's privacy.

Namaste [38]

Captain Endicott advised me to talk very little and listen a lot. He taught me two expressions and one mannerism that would help me be welcomed into to most networks and conversations while in Calcutta. I later found them to be in common use by millions around the world and magical for me many times.

As a polite, mannerly, and respectful greeting or farewell, the Hindus say the word, "Namaste, "(nahm-ahs-tay or num-us-teh) with a slight bow and hands pressed together with thumbs at chest."[37]

Islam forbids Muslims bowing to any person—only to Allah. They commonly use an expression of greeting, farewell, or appreciation; they say, "Asslamualaikum" (Ah slahm mu lay kum) meaning may peace be upon you."[38] In time, I used those two expressions and jestures to my advantage around the world and was embraced because of it. Subsequently, I learned new ones that connected me similarly with other countries.

I continued with wonderment throughout the market area and never in my life have seen such crowds of people walking and

Market [39]

47

shopping among the street vendors. Monkeys and cows by the hundreds were everywhere, going in and out of shops at will. I was careful where I stepped and got over being startled by the frequent tugging at my pant legs by the monkeys.

It was a bit unnerving every time I heard a strange bagpipe sound; nearby would be a "charmer" playing a flute as a large King Cobra with head erect and neck spread rose out of a wicker basket; it would sway in unison with the movement of the charmer's flute. These snakes grow to about 15 feet in length and are highly venomous although I learned most have been defanged.[39]

Snakes are typically revered in the Hindu religion as are all creatures. However, there are over 200 species in India of which 60 are highly poisonous. Several thousand deaths a year occur there from their bites.

Rattlesnakes were fairly prolific when the early settlers arrived in the Massachusetts Bay; however, I rarely saw a snake of any kind while growing up. I easily saw more in number and species in baskets on the streets of Calcutta my first day than I had seen in my entire life.

Snake Charmers [40]

Snake Charmers

I continue to be reminded of how much religion or lack of it is at the root of all our thoughts, words, and deeds, regardless of the time or country of origin.

The reverence for all life and a belief in reincarnation are basic doctrines of Hinduism, the dominant religion of India. This explains why even animals and reptiles are considered to be sacred and are never to be harmed. It is thought that after death, there is a rebirth of the soul—transmigration—to a different body or form based on how they lived their last life. This includes moving up or down the scale of classes as well as from non-human to human forms—animals, insects, and reptiles to human beings or vice versa.

"Hinduism is considered to be the world's oldest religion dating back to 2,000 B.C. Nine of every ten of that faith live in India. There are hundreds of millions of followers worldwide—the third most behind Christianity and Islam. Hindus and Muslims have been at odds back to the times when India was under Mughal domination. Centuries earlier, approximately 100,000 so-called infidel Hindus were murdered in one day over religious differences.[1]

"Islam continues to maintain substantial representation in Calcutta and is the fastest growing. Although I was reared in the Christian faith, I have grown to accept that some of the principles of Hinduism are consistent with my Christian beliefs, particularly regarding the sanctity of all life and the worth and dignity of all creatures.

"India is the world's largest democracy; however, the belief in the caste system somewhat tied to Hinduism, ancient law, and tradition seems to be anything but democratic. There are four levels of rank according to perceived capability and or birthright. The highest and most favored are the teachers and priests. Somewhat lower are the warriors and rulers. Further down the scale are farmers, traders, and merchants. On the bottom are the laborers. A fifth category exists but lacks the dignity of an official caste ranking. They are the outcasts who are street sweepers and latrine cleaners. Known as 'Dalits,' they are considered unclean for a number of reasons. Included in this unfavorable category are all the lepers and over ninety percent of the poor and illiterate, together totaling in the millions across the country.

"Leprosy is a hideous bacterial disease highly peculiar to India, dating back to 2000 B.C. The Bible mentions it many times, including in Luke 17:11-19, with the miraculous healing of ten lepers by Jesus. If left untreated in the early stages, it causes severe physical deformities and the victims to be scorned and ostracized to the streets or into colonies.[2]

"I came upon many lepers along the roadsides and alleys with crippled limbs, no toes or fingers—some begging—but most pleading in silence with pitiful looks and sunken eyes, yearning to be rescued. I was haunted for years

---

[1.] This strife was resolved to a degree as revealed in a look-forward to 1947. British India was divided into two independent nations basically according to religious affiliation—India for Hindus and Pakistan for Muslims.

[2.] A glimpse into the future reveals that it became treatable but strangely remained as a plague throughout India.

by the memory of those hollow expressions of despair; I had to walk away, helpless to do anything about their plight. I only saw that look once again in my lifetime. It awakened the horrors of the past. I seemed to hear their groans again pleading, 'please help us, please.' God gave me another chance."[40] I will tell you about it in time.[1]

Captain Endicott has alerted us that we will be heading home within two weeks. I had longed to visit other places in India, but distance and time made it unworkable. Maybe in the future, I can travel to the Eastern Border to the Himalayan Mountains. Everest is the tallest peak in the world and is a part of that chain; maybe I could see a Yeti, the Abominable Snowman. Alexander the Great in 326 B.C. demanded to see one when he conquered the Indus Valley.

The Taj Mahal at New Delhi is another site I yearn to visit. It is one of the seven great wonders of the world and represents one of the most incredible love stories of all times.

The Ganges is the third largest river in the world and is considered sacred within the Hindu religion. I would enjoy following its rich history as it flows through all the great cities from the Himalayas to the Bay of Bengal. There is so much to see, to learn, to do— so little time.

> " Rescue the perishing, care for the dying... duty demands it; strength for thy labor the Lord will provide..."[41] —Fanny Crosby

In retrospect, this voyage prepared me well for just about everything there is to know about sailing and commerce. By the way, I finally learned that our original cargo was cotton and tobacco. Captain Endicott made numerous buys, sales, and trades for handsome profits for the ship owners and investors, along with a nice bonus for himself. We brought back black pepper and tea, lots of it. Amazingly, it was sold for an enormous profit within days of our returning to Salem Harbor; it exceeded the entire original cost to build our ship!

Second Voyage to Calcutta

The enormous success of this voyage prompted a return trip to Calcutta in the coming months. It equally was profitable, especially for me. My job title remained as Cabin Boy, but my job description changed substantially. Captain Endicott allowed me to serve as a Relief Mate substituting for the regular

---

[1.] Queen Pomare on page 150 gives me the second chance.

seamen when they were ill or needed rest. I was able to get direct experience in every category of seamanship. My confidence was bursting, and my attitude was on a high, except for one missing link.

It was an especially momentous day in my life—that last day when we would be docking back home again in Salem Harbor. I was on deck doing my chores when Captain Endicott from the helm called me to his side. Still in my teens, I was startled by his words. It is strange to me how words can affect emotions. They are only letters configured in different ways, yet in certain arrangements, they have the power to change feelings instantly. They can bring joy, sadness, and even anger with equal indifference; they can sting, soothe, or cause retaliation. Accordingly, Captain Endicott said to me, "William, you are ready to fly the nest; it is my duty and honor at this moment to certify you as an official Mate so take the wheel and bring us in."

Tradition suggests that giving me complete control of the *George* at that moment was the symbolic equivalent of granting a battlefield promotion. In the words of medieval guilds, I had served on previous voyages as a Cabin Boy: on a subsequent voyage I was given more training and responsibility in the role as a Journeyman—with an increase in pay. This reclassification to the status of Mate assured captains throughout the area that I had demonstrated competence and was ready and able to serve as second in command.

"In the fell clutch of circumstance
  I have not winced nor cried aloud…
  It matters not how strait the gate,
  how charged with punishments the scroll,
  I am the master of my fate,
  I am the Captain of my soul".[42]

Invictus—William Ernest Henley

I was not home for long until I received some very good news. I was offered and accepted a job to be the First Mate on the *Jason*. Its ultimate destination was to Gibraltar via Havana, Cuba.

CHAPTER 5

My First Voyage as a Mate

Yasmina Bounty [41]

"Twelve ships of the British Royal Navy have been named after Jason, a Greek mythological character. As legend goes, he was considered to be a hero after he carried a distressed old lady on his back across a stream. Later, he needed a ship for an expedition and gathered other heroes, including Hercules, to accompany him.

A View of *Ye Jason Privateer*[42]

They set sail and experienced a series of morbid and sordid sea adventures that ended with Jason being killed by a falling piece of wood from his ship. Subsequently, he and his ship were raised to heaven."[43]

Jason's connectivity with the sea perhaps explains its popularity in naming ships. I was very curious about which one of several *Jasons* would be the ship for my maiden voyage as an officer.

My inquiries led me to some prints of very old watercolor drawings by Nicholas Pocock. He was an apprentice on the above ship in about 1760. This would make this particular *Jason* at least 60 years old by the time I took my first voyage as a Mate. That age for a ship was not unusual because of the special wood, materials, construction, and maintenance during this era. There are numerous floatable boats in museums that are over one-thousand years old.

"Pocock identified this *Jason* as being French-built, which fits the time frame of this narrative. The ship's crew is shown in the illustration above preparing to set sail for Africa as the pilot and commander arrive in small boats, apparently taking captives on board. This *Jason* made at least one confirmed 'slaving' voyage for Edward Willcocks & Co. to the West Indies. According to Pocock's account, it delivered 340 enslaved Africans from Angola to be sold in Jamaica. Captain John Clark had orders to buy 600 enslaved Africans, but it is not known how many he purchased or how many died on the transatlantic crossing." [44]

Some eerie coincidences connect me to this beautiful island—though infamous for its dreadful past. Slavery and sugar cane were imported by the Spanish; breadfruit plants came from the South Pacific on a British ship command-ed by the legendary William Bligh; and I, William Driver, soon will depart Salem on what appears to be the former slave ship which no doubt had made many voyages to our same destination—loaded with humans in bondage. This damnable practice will not have ended by the time we arrive in Jamaica.[1] How-ever, this *Jason* will not dock with William Driver on board if our cargo is to include slaves, going or coming.

I have concluded my research and have no doubts that this particular *Jason* is the same French privateer slave ship that was captured in 1813 by the *HMS Venerable*. Privateer ships were owned by individuals or investor groups to function as mercenaries. They were very good at providing services for fees in return for protection or aggression; but if captured, they became the prop-erty of the victors. Subsequently, this *Jason* indeed was captured by American privateers during the War of 1812.

The American Navy (Merchant Marine) captured only 254 enemy ships during this War while the American privateers captured 1300 British ships valued at $40,000,000. An incredible song was penned two years later by Fran-cis Scott Key, who watched one of the final battles at Fort McHenry, Maryland, i.e., *The Star-Spangled Banner.*[45]

Ultimately, this *Jason* ended up in the hands of an American owner or a group of investors ending a convoluted series of events that worked very much to my advantage. I recently was promoted to Mate status while on the *George* serving Captain Endicott. Fortunately, I will serve as his Mate on the pending voyage on the *Jason*.

I pause again in appreciation of Professor Hacker to let it be known that I faithfully am following his advice earnestly preparing for every tomorrow. I have been studying Spanish, which is spoken in Cuba and in sections of Gi-braltar—my forthcoming destinations. The basic Italian phrases I learned in Leghorn and the Bengali in Calcutta convinced me of the tremendous business advantages I gained by being even slightly bilingual in non-English speaking countries. As well, I am learning from the ship's library resources as much as I can about their culture, history, religion, and principal commodities we likely will be trading.

I well remember how devastating and uninformed a slip of the tongue or offensive gesture could be to a business relationship. I do not want something unwittingly to cause damage to a relationship or ruin a trade. So, when we ar-rive in Havana and Gibraltar, I will be well-informed.

---

[1] Historically, the island of Jamaica was claimed by Spain when Columbus landed there—not on the mainland of North America—in 1492. It subsequently became a British Colony in the early 1700s and remained such until 1962 when it was granted its independence.

The Capture of Havana[43]

Havana, Cuba

The sea was at peace with the world that morning as the fog was lifting in sequence with the "land ho" alert of our ship's watch. I was in a reflective mood. We were miles away, but I never have seen a more impressive site as that perched there high on a massive rock bluff overlooking the entrance to the harbor of Havana. My research had prepared me for the significance of Morro Castle but not for the amazement. It originally was built in 1589 to protect the harbor. Later, an added feature of protection was a massive chain strung across the mouth which could be raised and lowered if suspicious ships were sighted. The British in their on-going quest for new territory captured the fortress by a land attack in 1762, but it was returned to Spain, the current ruler, a year later.[46]

The wharf soon will be coming into view, and all hands will gather on deck preparing to moor. The *Jason* is at a slow drift as we await the harbor master's permission to proceed. My thoughts are reminders of the mental preparation necessary for any merchant ship's Captain, Mate, and Trading Officer immediately after docking at a new destination. The focus immediately must be on the relevant country's exports and imports; it is different at every port and the minds must be refreshed accordingly. Both the hosts and the guests need to sell or trade certain commodities and products as well as to buy or trade for others. The basic law of supply and demand creates a balance that must be achieved for economic survival. Therefore, to understand what is in demand in Cuba

Havana Wharf [44]

needs to be factored against what it has to supply. Perhaps it is interesting to know how we obtained this information in advance—the essential question.

We are at a time in my history where communications between the seafaring are transmitted by word of mouth from one seaport to the next. The inns, pubs, and taverns serve as gathering places and serve as the primary conduits for information to flow and be exchanged. There, the merchants learn from one another the news about international commerce—what to buy and what to sell. Accordingly, we are bringing from Jamaica cocoa, indigo, yams, and coffee, based on our most recent information that they are in shortage on the island of Cuba. Informed decisions and skilled trading give us our best chance of making this a successful business venture for the owners and investors who hired us.

I learned another important truth from Captain Endicott. It is much more important to determine what people "want" than what they "need"—a strong tendency in human nature. Consequently, we are going to Cuba to load up on two of the world's greatest wants, i.e., Cuban cigars and their rum. Both will be high priority wants for the European traders when we reach Gibraltar.

Cuban tobacco is unique in the world. It is not the kind of tobacco widely known back home in the southern states. It is unusual because of a rare and distinctive combination of seed, soil, and conditions that are unparalleled in the world. Soon after his discovery of Cuba in 1492, Columbus saw the natives inhaling smoke through their noses from smoldering tobacco leaves. They gave him samples to take back to Spain which helped spread its use throughout Europe during the 16th Century.

Cuban Hand-Rolled Cigar Artemiseña Factory [45]

Its popularity increased rapidly because of its addictive properties, along with its use in ceremonies, religious rites, and for medicinal purposes.

Connoisseurs rank cigars made from this exclusive Cuban tobacco to be the undisputed best that "big" money can buy in terms of mildness and taste. The supply rarely equals the demand, which drives the price out of reach for ordinary incomes. Conversely, the wealthy will pay exorbitant prices for them.

Water-Powered Rum Distillery [46]

Many a deadlocked business deal has been consummated under the spell of prime Cuban cigars. We plan to parlay our investment in these prized products many times over upon arrival at Gibraltar.[47]

Cuba's other esteemed product, which we believe will be highly sought on our arrival in Europe, is its renowned rum. We will find it in abundance in Havana, having been aged and mellowed for many years in hogsheads. When we arrive at Gibraltar loaded with cigars and rum, we will be highly welcomed and will find ourselves to be in a powerful bartering position.

This is my first trip to Cuba, so Captain Endicott encouraged me to follow the same protocol to orientate myself as before. My Spanish is quite good now from previous studies. Therefore, I have focused my leisure time on becoming knowledgeable about the country's culture, economics, history, and religion.[48]

I learned that various indigenous cultures had occupied the island for times immeasurable when Columbus arrived and claimed it as a Spanish Colony.[1]

Although we were at a peak time in slave trading and would be arriving on a former slave ship, the owners and crew of the *Jason* hate the practice and are

---

[1] Cuba remained one of the most common destinations for slave ships until the practice was abolished in 1886. After several liberation wars, Cuba ultimately gained its independence from Spain at the end of the Spanish-American War. More than a million slaves were brought to the island throughout its history primarily to work on the sugar cane plantations making it the world's primary producer of sugar.

"Slaves Cutting the Sugar Cane" [47]

committed to never participate or promote it in any way, form, or fashion.

I found that religion in Cuba is rooted in Roman Catholicism along with the spiritual traditions brought from Western Africa by the slaves.[1]

We were in Havana for a short while and completed our trades very satisfactorily. I only had a day to site visit which led me to a cigar factory, a sug-

Columbus Cathedral [48]

arcane mill, a rum distillery, Morro Castle, and the Columbus Cathedral where the explorer's remains are thought to have lain in state waiting to be returned to Seville, Spain for his burial.

Tomorrow we will depart for Gibraltar with the ship's hold fully loaded with premium cigars and rum.

En Route to Gibraltar

The *Jason* is slicing slowly through choppy water this morning as all hands are on deck completing their departure tasks for our voyage to Gibraltar. As the Mate, my job is to oversee all work stations and report the status to Captain Endicott. He is in his quarters reviewing the charts in preparation for this 4,000 nautical mile crossing of the Atlantic. There are play-

Albatross [49]

ful dolphins, colorful trogons, and a variety of gulls escorting us as we approach the mouth of Havana Harbor. An albatross is gliding gracefully above the other birds seemingly beckoning us to the open sea—a good omen!

---

[1] It remained a deeply religious country until a glimpse into the future reveals that in 1959, Marxist Fidel Castro became the Prime Minister as the result of a major revolution. He essentially thwarted freedom of religion by closing churches and over 400 catholic schools. One could not be a member of the controlling Communist Party unless they professed atheism. He died in 2016 and conditions appear to be improving.

If you did not know it, sailors are a curious lot who believe that seeing one of these majestic birds in flight brings good luck, but killing one is a certain curse foretelling pending doom. Perhaps our superstition comes from working in such a hazardous profession where so many things can go wrong. In fact, mariners have the second-highest mortality rate of any occupation except for loggers.[49] I know it

> **"FOR GOD HATH NOT GIVEN US THE SPIRIT OF FEAR; BUT OF POWER, AND OF LOVE, AND OF A SOUND MIND."[50]**
>
> —2 TIMOTHY 1:7

well; the sea has claimed two of my grandparents and made my parents fearful for me to follow their path. I respect their concern but do not share their fear; my goal to become a Captain is unchangeable and relentless.

After we were underway a full hour, Captain Endicott came from his quarters and asked me to follow him to the wheel. He relieved the helmsman and asked us to study the chart and notes he brought for us to become familiar with the navigational plans he had prepared. We went over every aspect, and both felt confident we understood his decisions and rationale for them.

Thereafter, he released the wheel back to the helmsman and called me aside to discuss other plans he had in mind. He shared that, regardless of me still being a youth, he felt I was becoming close to qualifying to be a Captain except for in one domain—navigation. He was not comfortable if he became incapacitated, that I was ready to assume full command. I fully agreed; there were many terms, processes, and procedures where my understanding was shallow and my confidence inadequate.[51] I was delighted when he said he would like to begin focusing on them to better prepare me. This leg of our trip was estimated to take several weeks, so there would be a lot of free time to devote to strengthening my knowledge and skills. I expressed my deep appreciation for his offer and asked when we might begin with his mentoring. He smiled and responded that we would start immediately with a review of the basics; he asked that I follow him to the stern of the *Jason*.

The Captain picked up one of the ship's waterproof slates, and with a soft stone pencil, drew a pyramid with letters similar to this adjacent image. He stated that the D = distance, the S = speed, and the T = time. If I were to place my finger over the factor I wanted to determine, I

Distance-Speed-Time [50]

could multiply or divide the other two and produce the answer. The horizontal line represents the D (distance) being divided by the appropriate choice below the line (either speed or time). Covering the S requires the D (distance) to be divided by the T (time.) Covering the T (time) requires the D (distance) to be divided by the S (speed.) Covering the D (distance) requires multiplying the S (speed) times the T (time). All piloting and maneuvering solutions at sea contain these three variables; therefore, understanding how to calculate them is fundamental to all navigation decisions.

Captain Endicott was somewhat like Professor Hacker in his teaching techniques. He would explain the principle, give examples of a practical application, and then provide an opportunity to practice. Finally, he required proof from me that I comprehended by demonstrating what he had taught me.

His most practical example was to determine how long it would take us to cross the Atlantic from Havana to Gibraltar. He wrote on the slate the one known variable we determined from his navigational chart, e.g., the distance being 4,000 nautical miles (4,600 land miles). He calculated that our speed would average about 6 knots (6.9 miles per hour). He clarified the distinction between nautical terms and statute (land) terms which initially was very confusing to me. Nautical miles are based on the circumference of the earth and are used exclusively when charting and navigating. A nautical mile is slightly more than a land mile—15/100 more. It takes 1.15 land miles to equal a 1.0 nautical mile.

The Log-line

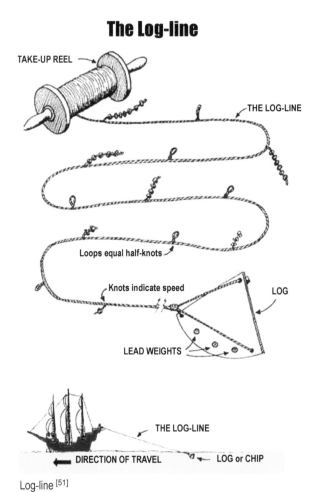

TAKE-UP REEL

THE LOG-LINE

Loops equal half-knots

Knots indicate speed

LOG

LEAD WEIGHTS

THE LOG-LINE

DIRECTION OF TRAVEL   LOG or CHIP

Log-line [51]

Captain Endicott said the word "mile" being used for both land and sea with two different meanings is what causes the confusion; it takes a while for one converting between them to become comfortable. He explained there are some very practical reasons why mariners worldwide use this system for navigation, which will be a future lesson for me in greater depth.

Nevertheless, we found our answer by dividing D by S or 4,000 ÷ 6=666.7 hours, which is about 28 days (666.7÷24)—the time it likely will take us to arrive at Gibraltar. As a rule of thumb, roughly seven nautical miles equals eight statute miles; therefore, we can convert nautical to statute by multiplying nautical miles by eight and dividing the product by seven.

I was able to solve several problems determining distance, speed, and time. Later, I asked why the usual terms for measuring speed in miles per hour would not be easier than using knots. My answer came in the form of another demonstration. He took from a storage box a device he called a "common log." It appeared to be a long coil of rope with knots tied every 47 feet and 3 inches apart. The exact separation of the knots had been determined mathematically and proven by years of experimentation. The terminal end of the rope was tied to a pie-shaped weighted piece of wood. Accompanying the common log was a 28-second sand-filled glass—a mini-version of an hourglass. The procedure began by flipping the glass and releasing overboard the "common log" simul-taneously. When the last grain of sand dropped, the rope was stopped, and the knots drawn into the sea were counted. In his demonstration, we counted 8 knots (8 nautical miles per hour) which we divided into the 4,000 nautical mile distance telling us it would take approximately 500 hours to reach Gibraltar. This procedure was done every hour or so for refinement and recorded in the ship's log.

> "Deep in the wellspring of simplicity is wisdom. To take what is complex and make it simple again takes real presence and understanding." [51]

I was amazed at what this session taught me. Two of the four variables used in piloting and ma-neuvering at sea can be fairly closely approximat-ed by using a sand glass to measure the time and a knotted rope for speed. The third variable—dis-tance—is predetermined before we leave the dock.

I was about to become overconfident at its simplicity until Captain Endicott reminded me that we only had been talking about the most elementary factors. There is a more formidable fourth variable—direction—we would learn about in the next session.

The next few days were fairly uneventful as we sailed forth maintaining our projected course and speed. In my spare time, I followed the usual protocol in terms of familiarizing myself with the history, culture, economics, geography, and religion of Gibraltar.

It appears to be the most unusual place I ever will have visited—having only a 2.6 square mile area including a city and rock by the same name; yet, I won-

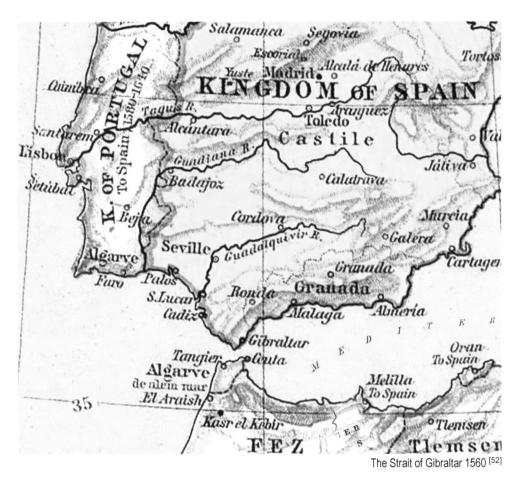

The Strait of Gibraltar 1560 [52]

der why such a small seemingly insignificant spot seems to be such a mecca for merchant ships.

Obviously, the Rock of Gibraltar is very strategic militarily as it guards the entrance to the Mediterranean Sea and separates its Strait to the south from Morocco in North Africa—another continent. To the north, it borders Spain with which a somewhat contentious relationship has existed for centuries. There is easy access by water to the east, south, and west, so its location is ideal for sure, but Captain Endicott suggests there is something else I have not discovered and gives me a clue. Mysteriously, he says I can find my answer in the sand. Now, he really has stirred my curiosity; in the sand…the sand…the sand, I mused? "Do not tell me unless I give up!" I rushed to the resource room to probe for the answer.

My research revealed that beginning 100,000 years ago, the Neanderthals, Phoenicians, and Romans sequentially occupied the area, but in 711 A.D. the Moors (dark-skinned North Africans) captured and reined over Gibraltar and the entire Iberian Peninsula—Spain, Portugal, and parts of France—until 1462 when expelled by the Spanish. Gibraltar eventually came under British Sovereignty in 1713.[52]

My clue about "sand" was unproductive, so I returned to Captain Endicott for help. He smiled and began slowly filling in the pieces of the puzzle. Despite its size, it is a mecca for trading merchants because of its uniqueness in the world of commercial trading centers. It connects to all the European trading routes, but most significantly it is less than ten water miles away from the Continent of Africa—the world's largest trading expanse—a territory that could hold China, India, and the United States combined. It is this connectivity that brought Captain Endicott to school me about why Gibraltar particularly benefited from

Tariq ibn Ziyad [53]

the sand—the largest sand pile in the world covering 3.5 million square miles—the Sahara Dessert. He started with a history lesson revealing more about the Moors from North Africa. I listened intently.

He revealed that in the eighth century a Moor Commander, Tariq ibn-Ziyad, crossed the nine-mile Strait from Morocco and attacked Gibraltar which was under the control of Spain and an army of 100,000 soldiers. He saw fear and intimidation in the eyes of his men facing the overwhelming odds. They kept looking back at the ships as a means of escape. He removed that consideration by ordering the boats to be burned. His ultimate trust was in Allah. No alternative remained but to move forward into battle; consequently, they prevailed and helped introduce Islam to the peninsula.[53]

This example made me recall Professor Hacker's lecture on decision-making, where he admonished the practice of "second-guessing." He told us about Julius Caesar in 56 B.C. demonstrating how to strengthen courage and resolve; similarly, when he invaded Britain, he burned all of the boats, so there was no option to retreat. Tariq ibn-Ziyad's example at Gibraltar produced the same results although centuries later. Perhaps he had heard of Caesar's tactic. In 1519 during the Spanish conquest of Mexico, Commander Hernán Cortés burned his ships, leaving his men no choice but to conquer or die; they too prevailed.

I was struck by the fact that three of the greatest military leaders in history used the same psychological tactic to motivate and inspire their followers to victory. It appears to me these examples offer a practical application to all life decisions. I plan to draw from them in my decision-making in the future. I will analyze the facts, consider the options, choose the most logical, and not look back to second-guess my decisions.

Another historically significant leader followed Ziyad centuries later and

Mansa Musa [54]

helped continue the spread of Islam in every direction from his base in Mali, North Africa. This influence extended throughout the Iberian Peninsula and particularly in Gibraltar. His name is Mansa Musa, who was from near the fabled city of Timbuktu. It was well-known as a world-class cultural and educational center, but foremost as the terminal trading hub for all camel caravan routes crossing the Sahara—the end of the road.

It was considered so "far away" a common expression emerged in jest to describe any great distance, i.e., "from here to Timbuktu." However, it actually was not the end of the road as far as the major traders were concerned. Another thirteen hundred miles and a boat crossing were required to get to Gibraltar. Although Timbuktu was a significant trading center, it served as a stop-over station, inn, or oasis en route to the final destination where the really big trades took place—trades that in one journey could make a person wealthy for life. Gibraltar was another month away but was worth the effort. For example, there were times when an African merchant could sell his salt for an equivalent weight in gold.[54] Caravans as large as 12,000 camels crossed the Sahara each year which, at the expected forty miles a day pace, would take approximately three months. It seemed too far and time-consuming to me—on foot or riding a camel.

However, I learned another life lesson from the Captain's story. "Time and distance" are relative considerations in life. Much depends on where one is going and why. For example, a Pilgrimage from Timbuktu to Mecca in Saudi Arabia is a yearly event and is a mandatory religious duty at least once in the lifetime of every devout Muslim. It is 2,800 miles away—more than twice the distance to Gibraltar. Relatively speaking, it is understandable why the major traders were eager to take their wares on to Gibraltar; the reward was potentially much greater, and it took only a month or less to get there. Compared to three months crossing the Sahara, an additional month presented no major obstacle.

Captain Endicott apologized for leaving me in suspense about the main character in his story, Sultan Mansa Musa. He reconnected with him, revealing that he not only was the Emperor of the Mali Empire, he unmistakably was the King of Traders—perhaps the all-time king. He earned the distinction of becoming the richest man in the history of the world as a result of his exploits trading in gold and salt. There are accounts that he also was engaged in the slave trade.

It is of remarkable significance that numerous reports tell of Mansa Musa financing trading voyages across the Atlantic prior to that of Columbus. There

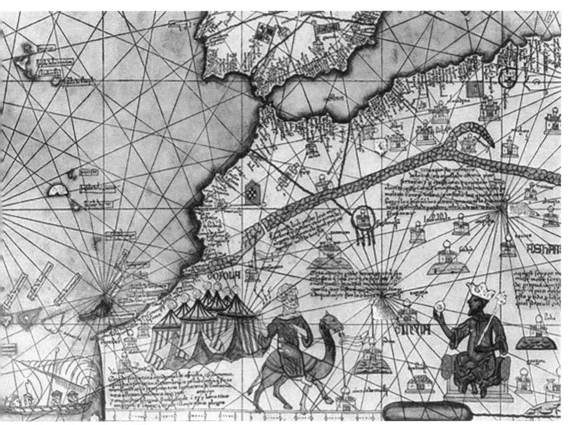

Mansa Musa's Domain [55]

is supporting evidence that they made it to the West Indies in North America. Columbus even noted in his journal that the natives confirmed to him that black-skinned people had preceded him by coming from the south and east in boats and trading in gold-tipped spears. Further, Columbus's son wrote about his father telling him that some of the people he saw in Honduras were almost black. It is most noteworthy that, in addition to other compelling evidence found in Central America, the very son of Columbus offers the strongest testimony that his father did not discover America; the evidence suggests it was an unidentified Black mariner from West Africa.[55]

Captain Endicott and I were wrapping up this very informative session on the significance of Gibraltar as a trading center when the ship's storekeeper interrupted to bring some disturbing news. The nature was such that the captain felt we should summon all the ship officers to his quarters for further discussion. After all had arrived, the storekeeper, with a somewhat distressed look, shared that we have an unanticipated growing shortage of water in the hold. The crew had been craving more than normal because of the prolonged high temperature and the high intake of salted beef and pork, but of greater significance was the discovery of a leak around the bunghole of one of the water barrels. It had lost all but a couple of gallons from the original thirty-one. Further, our rain barrels remained empty because we had not a single drop fall since leaving Havana.

Based on an anticipated additional ten days of sailing before our arrival, he calculated was that we need to take immediate precautionary action as follows:

- Ration each man's allotment from one gallon daily to one-half gallon.
- Assign two men to purify daily as much ocean water as possible.
- Remove one barrel of rum from storage and allot to each one tot (pint) a day.

After a brief discussion and unanimous agreement—particularly over the last recommendation—Captain Endicott asked us to pass the word for an afternoon muster point meeting where he will relay the problem, the remedy, and order for immediate implementation.

The crew knows about my promise to my mother that I never would drink alcohol, so several are offering to trade two pints of water for my one pint of rum. That consideration presents a moral dilemma for me. I want to give it

The Rime of the Ancient Mariner

. . . All in a hot and copper sky, The bloody Sun, at noon, Right up above the mast did stand, No bigger than the Moon. Day after day, day after day, We stuck, nor breath nor motion; As idle as a painted ship, Upon a painted ocean. Water, water, every where, And all the boards did shrink; Water, water, every where, Nor any drop to drink.[56]

—Samuel Taylor Coleridge

Water, Water, Everywhere [56]

more thought before making a decision involving a product my parents shunned and referred to as Demon Rum. Remarkably, everyone seemed to become more thirsty.

Captain Endicott and I reconvened to conclude our mentoring session. He summarized that Tariq ibn-Ziyad's success, along with that of other leaders who followed him, brought Spain under Muslim rule for more than 750 years until Queen Isabella and King Ferdinand completed the Reconquista (re-conquest) in 1492. Jews and Muslims by the thousands were expelled or forced to convert to Christianity across the entire Iberian Peninsula. Today, it remains predominantly Roman Catholic but recognizes religious freedom for all beliefs.

He drew a parallel to the time of my birth in 1803 when Napoleon Bonaparte's French Empire essentially ruled over all the landmass of Europe, and the British Navy controlled the seas. Gibraltar had been invaded, captured, defeated, and recaptured more than any place on earth—enduring 15 different sieges. The British, Dutch, French, and Spanish have been the major combatants. British sovereignty over Gibraltar was reaffirmed and continues as a result of the Battle of Trafalgar in 1805 when they defeated the combined French and Spanish forces.[57]

The Captain said we could expect to find a more tolerant multi-religious and cross-cultural society living in reasonable harmony when we arrive. However, we would be engaging with traders from all over the world and should be extra careful not to unwittingly offend any. My list of offensive prohibitions continues to grow as he and the trading officer gave me three more. I learned to never blow my nose in public, smell food, or to use my left hand to point, wave, or accept something. Captain Endicott suggested that I always ask a friendly contact in each new country to warn me about the areas of contention among their cultures.

His parting remark was a commendation of my efforts to become broadly informed about the people and places where our trading will occur. He said it may seem like unnecessary effort and unrelated to the business of trading commodities, but time will prove to me its value. He said that "knowing the territory" is equally as important as knowing how to navigate.

As I consider what I already learned from our stops in Jamaica and Cuba, and what awaits me in Gibraltar, this has to be one of the most enlightening experiences a person of any age could ever imagine. I pinch myself to affirm I am not dreaming. Still a young man, I already have been exposed to more cultures, economies, geographies, languages, religions, teachings, and brilliant minds than most of my former classmates combined will experience in their lifetimes.

Somehow I sense a premonition that God has a special mission and purpose for me and my life. Far too many good things are falling my way just to be a coincidence or luck. Strangely, there still is a void—emptiness I cannot explain...

Land Ho—Arrival in Gibraltar

Fortunately, we are reaching our destination without a further incident since our water shortage was addressed. It will be replenished with some to spare before our return trip to Salem.

The lookout already has proclaimed the sighting of the massive Rock off in the distance—an unmistakable wonder of the world at this unusual confluence of the Atlantic Ocean and the Mediterranean Sea. Unquestionably, we have arrived at a special place in the world for merchants.

In addition to the other advantages of this unique location, the cost of trading within the Port of Gibraltar makes it even more appealing. In 1706, Queen Anne made it a "free port" enabling international ships to conduct business without paying any type of duty, fee, tariff, or tax. This helps explain the wide variety of flags and the large number of ships we see docked and anchored in the bay. The ensign of each country is required to be displayed while in port; ours now has 23 stars since Alabama and Maine have been added. Captain Endicott declared that this requirement of citizenry gives us clues as to what might be on board other vessels. For example, a ship from India nearly always will carry an abundance of spices.

When we approached the docking area, the deck crew lowered our utility boat and rowed Captain Endicott to the Port Master's Shack. This customary procedure is for the purpose of obtaining permission to anchor and to file documents required by Maritime Law and those of the local jurisdiction. Additionally, the Port Master posts a list of each ship's location in the harbor and its cargo. The news spreads quickly in terms of what is available and of particular interest, the high demand items.

The Captain returned to inform us the dockside was unavailable, so we would spend our stay operating from an anchored position and tied to a numbered mooring buoy which is anchored itself. Our location is ideal in a wind-shielded cove. Since most of the trading will occur from ship to ship, we and others will use our utility boats to taxi to and fro; as well, we will be well-positioned to conduct further business in the markets ashore.

Within hours of our arrival, traders from three continents rowed by using megaphones to converse with Captain Endicott. If interest was shown in our wares, they would come aboard and discuss a possible trade. The news spread quickly that we had Cuban cigars and rum; the interest was strong and widespread. Appointments were strictly "first come-first served." As a newly designated Mate, I was invited to join the Captain and the Trading Officer in all discussions and transactions.

There is the usual tavern near dockside for the off duty crew to find food, drink, and frivolity. Checkers, dart throwing, and dice games are available along with a couple of local musicians who normally play traditional songs of the sea. I learned that the taverns are the most popular places onshore for captains to share and gather information; as well, they are convenient places to conduct trading.

Saturday Night at Sea [57]

The sailor's life is bold and free;
His home is on the rolling sea;
And never heart more true or brave,
Than his who launches on the wave;
Afar he speeds in distant climes to roam,
With jocund song, he rides the sparkling foam.[58]

—Sailing, Sailing—Godfrey Marks

Captain Endicott, not only was a skilled navigator, he was a smart trader. He held one of the most advantageous hands in the bay on this trip because of our goods. The Cuban cigars and rum were in such demand that in several early contacts, European traders, from their utility boats, attempted to negotiate for our entire store. I thought that really would be a good move; we could sell it all and head home! The Captain thought differently. His response to a similar proposal during the first appointment was very calculated; to be considerate to all the other buyers in light of our limited supply, he had decided to ration the amount he would sell in a single purchase. The potential buyer immediately raised his offer to a ridiculous level if he could buy us out. The Captain felt he needed to hold back in reserve some of the goods to leverage against purchases of other items to take back to Salem. Therefore, after some staged thought, he graciously "gave in" and raised his limit to two ration lots at the escalated price and the right to buy more if any remained before we were to depart. The entire supply was sold or traded within the first month at double Captain's original price—a price actually set by the first buyer.

I learned two valuable lessons that day about human nature: The influence of the law of supply and demand over decision-making is enormous and that greed can completely dull one's sense of reason.

> HE WHO IS NOT CONTENT WITH WHAT HE HAS WOULD NOT BE CONTENT WITH WHAT HE WOULD LIKE TO HAVE.[59]
> —SOCRATES

The rum went quickly in a similar fashion. We had emptied our ship's hold and reloaded it with equally valuable cargo, including a variety of spices, textiles, and French wine. There was some space remaining, but the Captain did not want to completely fill up just yet; he had another plan in mind.

Inclement weather moved in and delayed our departure for another day, so those who wished to go ashore were given permission. Most went to the tavern, but two of us decided to tour the city and get some culture. While well-rested, we walked a challenging trail to the top of the Rock. It was not steep like a mountain but inclined such as to make it very tiring. We saw several old folks with distressed looks resting on rocks along the way. Some had given up and were on their way back down muttering choice profanities as they descended. The view seemed to be from the top of the world and was breathtaking. We could see the continent of Africa in one direction, the country of Spain in another, and Gibraltar all around us.

Barbary Macaques Apes [58]

On our way down we saw tail-less Barbary Macaque Monkeys by the dozens. The signs told us they are the only wild monkeys on the entire continent of Europe. Further, there are over 200 caves including the most impressive St. Michaels, which is over 700 feet deep.

The rich history of the

Moorish Castle [59]

Moors was our focus when we visited their castle built in the 8th century. They occupied and ruled Gibraltar for over 700 years. To enter a structure built over one-thousand years ago made an unforgettable impression on both of us. We could visualize the battles taking place throughout the centuries.

We had enough energy for a final stop. We wanted to set foot in the country of Spain just to say we had been there. Additionally, it would give me the chance to practice my Spanish I had been studying faithfully. There were guards from both countries at a checkpoint station on the border. We had to obtain a permit just to walk across. Equally, the same was required of those wanting to cross into Gibraltar's territory. My first impression was that neither of the representatives was happy to welcome visitors. I soon learned that Spain thought that Gibraltar rightfully belonged to them, and the British had no intention of returning it. It was evidence of long-continuing disputes, wars, and bad blood. We did not enjoy the hard looks and cold-shoulders, so we left.

Overall, it was a great day, but we were worn out from the Gibraltar hike. We returned to the ship to get rested for tomorrow's departure.

Side Trip to Ceuta

With no specific time frame to be back in Salem, Captain Endicott announced that we are going to add a one day stop in close by Ceuta, Spain before beginning our return home voyage. This explains why he did not want to fill the ship's hold in Gibraltar. He reserved that space for three premium products in extremely high demand back home on the East Coast.

Ceuta is a popular and convenient stop for ships coming to Gibraltar because it is less than 12 nautical miles across the Strait and likewise is a tax-free port. Ceuta belongs to Spain even though it is located in a different country and continent—Morocco, North Africa. It is a popular trading center for the caravans having traversed the Sahara and desiring not to cross the Strait. Most importantly, it is the epicenter for the production of Moroccan leather which is considered to be the finest in the world; also, argan oil, which is native to Morocco, and is the rarest beauty elixir for the face and hair is processed there; as well, we will find the scarce Moroccan cork, which comes from the bark of a tree only found in the nearby Mediterranean Basin.[60] Captain Endicott shared his plans with the navigator and me before alerting the crew. Further, he informed me about how the three items came to be so rare and consequently so much in demand.

Leather Dyeing Vats [60]

Climbing Goats in Argan Tree [61]

Moroccan leather primarily comes from the skin of goats which are in abundance in the area. It is very soft and pliable and is in demand to make gloves, shoes, wallets, and to cover fine books. The leather cobblers back home are always in the market for it.

Their traditional processing uses sumac to tan and dye, along with a highly skilled and elaborate technique that highlights the grain. It nearly always is dyed red or black, but green, brown, and other colors are available. It is bought in whole or half hides or by the yard. Scrap pieces are marketable as well.

The source of the leather and the argan oil are interrelated. There are cloven-footed goats seemingly everywhere there are argan trees. They climb the trees and eat the fruit. It is unsettled whether they defecate the seeds or spit them out to grow more trees. Regardless, a traditional method believed to enhance the quality of the oil is to collect the manure, extract the undigested

Traditional Method of Making Argan Oil [62]

seeds, and manually process them to produce premium mixtures and higher prices. The argan trees thrive in the semi-desert areas of Morocco where climate and soil are so challenging that no other kind of tree will survive. Strangely, it takes 50 years for one to produce fruit, but they often live to be 250 years old. The fruit is referenced by the merchants as the gold that grows on trees. It is very labor-intensive and takes pounds to extract a single ounce of the oil—the most expensive vegetable oil in the world.

The relationship between the goats, the trees, the fruit, and the leather sends a compelling testimony that makes me pause. Such synergy is impossible to happen by accident or coincidence without the hand of God—Muslims would say the hand of Allah. Since with most Muslims, Allah is the only God. Logically, it has to be the same God worshiped by Christians and Jews. They as well believe there is only one God.

Some of the finest cork to be found in the world comes from the bark of the cork oak trees common to North Africa. Its uniqueness is best revealed by the fact that there are 800 million cells found in a single bottle stopper. The spongy product is stripped from the trees like shearing sheep and is stored in planks to dry. It regrows and can be stripped again about every nine years. Cork has hundreds of uses, such as for flotation insulation, packaging, padding, and wine bottle stoppers. A single mature cork oak can produce as many as a hundred thousand cork stoppers.[61]

The market place in Ceuta was loaded with fine Moroccan leather of all

Bark of Cork Tree and cross section of Cork Oak [63]

sizes, shapes, and colors. Argan oil was available although at shocking premium prices. Fortunately, we found the highest quality of cork to be in surplus supply, which enabled us to purchase it at bargain prices. Since Ceuta is a colony of Spain, I was able to use my knowledge of their language in helping Captain Endicott negotiate our trades.

We concluded our business fairly quickly, loaded our purchases of cork, leather, and argan oil on the *Jason*, and set sail for Salem. Captain Endicott remarked that our trip has been so successful that after all expenses, the owners will make enough money to purchase another ship—free and clear.

I was in awe looking back at Gibraltar to my left and Ceuta to my right as we were moving out of the mouth of the Mediterranean. Captain Endicott suddenly placed his hand on my shoulder and said he had a parting story for me. "In Greek Mythology, Hercules was the strongest man and greatest hero ever to have lived. Legend has it that he split a mountain in two with his sword and created the Strait of Gibraltar. The mountains on both sides remained as pillars to keep the sky from falling. Plato mentioned this place over two thousand years ago when he wrote that the mysterious lost city of Atlantis was located just beyond the Pillars of Hercules.

"Throughout the world, pillars are symbolized on buildings, money, statues, and endless other representations to indicate strength and stability. William, although this story was born in fiction and mythology, you needed to hear it because you now have actually been to the site. More so, it is a good allegory that encourages us to remain strong in upholding our principles, to be supportive of others in need, and to show stability in the face of crises."

The site of the Pillars of Hercules[62] faded in the distance from my view, but never from my memory as one of the most meaningful stories I had ever heard.

Pillars of Hercules [65]

CHAPTER 6

Antwerp, Gothenburg, and Liverpool

Seaport at Sunset [66]

We had been back in Salem two weeks after a pleasant 21-day crossing from Ceuta. I no sooner had my land legs back when I was contacted by Captain Isaac Chapman, asking me to be his Mate on a voyage back to Europe. He had heard of me through Captain Endicott as a possible replacement for his regular Mate who was recovering from an injury. He was scheduled to sail in a week on the *Batavia* to Gothenburg, Sweden with stops at Liverpool, England, and Antwerp, Belgium.

I asked for two days to give him my decision while I concluded some business matters in Salem. Additionally, I privately wanted to learn more about his reputation as a person and seaman. Captain Endicott knew very little since they only had met for the first time at the wharf's business office. I was able to meet the shipowners who were there inspecting another of their ships. They were impressive and encouraged me to sign on with them because of the potential for roles of greater responsibility in the future.

I spent the remainder of the day speaking with several who were familiar with Captain Chapman, including two who had sailed with him. I learned that he was 27 years my senior and had a son by his same name, three years younger than I, who wanted to become a Captain as well. However, my information about the father was marginal, except that he was a master navigator. He had mentored several Mates who had become Captains faster than normal, so that fact influenced me to agree to be his Mate under the condition he would school me in all areas of navigation—my greatest weakness. He agreed and assured me I would be well-prepared by the time we arrived in Liverpool.

I asked for more details about his plans and the nature of our cargo going and coming. Strangely, he was somewhat evasive. He was not sure if we would be taking products to trade and was uncertain what we would be bringing

First Trip to Europe [67]

back. It depended on the foreign markets and the prices at the times of our arrival and departure. We would be well-financed to take advantage of whatever favorable opportunities we might find. I found this unusual based on my previous experiences but passed it off as a different kind of Captain with a different approach and style. The opportunity for on-the-job training given by a master of the trade was sufficient motivation for the moment. Regardless, there was something mysterious that lingered unresolved in the back of my mind.

Our departure was still three days away, which gave me the opportunity to do some destination research in accord with my customary practice. I became broadly informed about Liverpool, Antwerp, and Gothenburg—the people, their cultures, and their economies.

Liverpool

I learned that the first enclosed wet dock in the world was constructed in Liverpool. It accommodates 100 ships and is unaffected by the tides and weather. Initially, the port traded heavily in coal, salt, sugar, textiles, and tobacco. However, less than two decades ago—at the time of my birth—ships from Liverpool had transported over 45,000 slaves from Africa. Slavery still has not been abolished in the British Empire, even though the trading of them was made illegal in 1807. Regardless, underground trafficking continues particularly within markets in the southern United States where the institution is widespread.[63] The very thought of human bondage sickened me the first day I was exposed to it, and my hatred of the practice grows stronger every day I live.

Captain Chapman continues withholding from me his plans for purchases on arrival, and I continue to remain suspicious.

Antwerp

Another concern arose while researching Antwerp, our second scheduled stop. Storm clouds of war are all over Belgium as a result of Napoleon Bonaparte's on-going attempts to become the dictator of the entire European continent. I was one year old when he started his conquest—not all that long ago—but five million people have been killed during that time frame. He already had been captured by allies and exiled to the island of Elba from which he escaped and gained control of France again. The allies responded likewise and defeated him at Waterloo in 1815 which is only 35 miles from Antwerp. Napoleon was exiled a second time and sent to the island of St. Helena where he remained until his death in 1821. As the saying goes, he met his Waterloo; but the entire region has been left unstable and in the recovery stage as a result of the twelve years of war. What possible attraction would draw us to Antwerp at this fragile time in their history?

I did learn that Antwerp has been prominent in the world diamond trade since the 15th century, but the constant upheaval apparently shifted much of that business to Amsterdam. I just will have to wait and see what Captain Chapman has in mind.

I noticed that they speak three languages in the area, e.g., Dutch, French, and German. I will learn some basic expressions in all three to add to my international background. There has been conversation back home of future voyage possibilities to the French Polynesian Islands, so I especially will focus on French just in case it is useful in my future.

As a side reflection, it seems so unlikely that Napoleon became such a successful leader. He acknowledged having a complex due to his small stature (5' 4") about which he was mocked frequently. He performed poorly in school and finished near the bottom of his class, and was from a disadvantaged poverty-stricken environment. Yet, he became feared and respected like few other men in history. The Duke of Wellington said his presence alone on the battlefield was the equivalent of 40,000 men.

I am a short fellow; I must continue to remind myself not to allow physical appearance to be a handicap, but to focus on qualities that would earn trust and respect in my potential in leadership roles.

Gothenburg

My research revealed several interesting facts about our last scheduled port, Gothenburg, Sweden—a major world trading center. It seems that the country has been in off-and-on conflicts for centuries with its neighbors in Denmark and Norway. It is strategically located as the country's only direct access to the North Sea and the Atlantic Ocean.

Iron, tobacco, wood, sugar, and herring fish are its main exports. It particularly is famous for operating Scandinavia's largest fish auction and as being a center for shipbuilding and repair.

Whatever awaits, I have committed to this voyage, and we are underway. We have had "fair winds and following seas this morning"[64]—a good beginning for this 2,700 nautical mile crossing. We anticipate attaining 100 nautical miles for each day's run (24 hours), so we should be in Liverpool in a month. After everything seemed to be going well, Captain Chapman called me aside to a storage box where we sat to have a chat.

He really meant business when he promised to get me up to grade in navigational knowledge and skills. We were barely out of sight of land when he handed me a chip log and asked me to demonstrate my capability of measuring our speed in knots. I was experienced at that and performed it to his satisfaction, confirming that we were moving satisfactorily at six knots. He asked if I had ever been lost at sea and if so to tell him about it.

I had not been lost at sea as such but shared my experience of being lost in a fog on Massachusetts Bay. A friend and I decided to row across one of the coves in his father's small fishing boat. We had done this many times in no more than

30 minutes. That morning, as soon as the shoreline was out of sight, a heavy fog dropped in on us like a blanket. I barely could see my friend much less the other bank. We kept correcting the direction we thought we needed to row because looking back at the wake appeared as though we were moving in a circle. We heard nothing and saw nothing other than a dead mallard duck floating in the water. After at least an hour of tiresome rowing, we were shocked to come upon that same duck again. Indeed, we were rowing in a circle and getting nowhere. It was a good two hours before the fog lifted, showing we were no more than fifty yards from where we began. Yes sir, if that experience counts, I have been lost at sea.

Captain Chapman chuckled and said it was a perfect example to illustrate the necessity of learning navigational aids when away from landmarks. "William, you needed to have used what I call CKN, Common Knowledge Navigation, in order to get you and your friend out of trouble. Had you been on the big water, they still might be looking for the two of you.

In the worst case of being lost at sea with no instruments, there are clues to help you. The sun, moon, planets, and stars along with the trade winds, currents, water temperature, birds, and whales were used by the ancient mariners, even to circumnavigate the entire world. They are our friends equally today. Had a Viking been in the fog predicament you found yourselves, he would have used a sunstone[65] to enable him to locate beams showing through the fog or

Viking Sunstone [68]

heavy clouds. Once the sun or any heavenly body is identified, a knowledgeable mariner can determine his general location and an appropriate heading.

Fortunately, current-day captains and navigators do not have to rely on the crude aids used by the ancients other than for emergencies. "William, every ship you sail upon in these times will be equipped with a nautical almanac, chronometer, clock, compass, chip-log, fathom sounding line, hourglass, magnifying glass, parallel rule, protractor, quadrant, sextant, sundial, telescope, dividers, and nautical charts. They appear to be simple devices, but their use still requires considerable knowledge and skill. I am confident that you have seen them utilized in your previous voyages. However, in case something was missed or overlooked, let's start with the very basics as though you are hearing about them for the first time."

The captain explained each instrument as to its purpose, function, and utilization. Later, he said he will allow me to practice with them and become

Ptolemy the Cartographer [69]

proficient. However, the first priority is for me to familiarize myself with every aspect of the nautical chart he handed me. He emphasized that I will have to pass a verbal or written examination demonstrating knowledge and adeptness with all the preceding before earning Captain status. I made notes on my slate as he spoke. I learned that Claudius Ptolemy is the father of astronomy and geography. Navigation charts are one of the fruits of his years of study and application.[66]

In a navigation context, a degree is a unit of measurement of an arc or angle. There are 360 degrees in a complete circle—any and all complete circles. Therefore, a single degree is 1/360th of a given circle. The mathematics likely originated with ancient Babylonian astronomers who used Base 60 rather than our Base 10 for computations. Remarkably unique, the number 360 is divisible evenly by 24 other numbers and is the smallest natural number divisible evenly by every natural number from 1 to 10 except 7. It is the perfect enabler for many things but especially for charting and navigating.

I observed that this degree wheel which we use to measure angles shows, as expected, 360 of these degree designation lines. Not shown is the fact that each degree contains 60 minutes, and each minute has 60 seconds. Therefore, there are 3,600 arc-seconds in a single degree—far more than can be displayed in this graphic. This is not to be confused with time although

Degree Wheel [70]

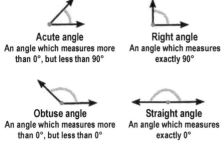

Acute angle
An angle which measures more than 0°, but less than 90°

Right angle
An angle which measures exactly 90°

Obtuse angle
An angle which measures more than 0°, but less than 0°

Straight angle
An angle which measures exactly 0°

Common Angles [71]

the terminology is the same. The divisions in this context relate to parts of an angle, not parts of clock minutes. Knowledge of angles and degrees of angles is the fundamental foundation of all navigation.

Now let us apply these elements to a type of map known as a navigation chart. Noticed that this image of planet earth is rotating with a 23.5-degree tilt from zero which has no tilt.

It is rotating eastward on its own axis—an imaginary line passing through the center from the North Pole to the South Pole. The earth rotates a complete circle of 360 degrees about every 24 hours. I said "about" because there is a four-minute factor involved when we bring the stars into the equation. For illustration purposes, we will use a rotation of 360° in 24 hours as our standard.

While the earth is rotating on its axis, it also is revolving around the sun at the same time; it takes 365.256 days for earth to orbit the sun one time. Again, for illustration purposes, we accept 365 days as our standard. In one year, the earth travels 940 million miles at a speed of 67,000 miles per hour to accomplish this feat. Next, we will flatten this globe and convert it into a special kind of map—a navigation chart.

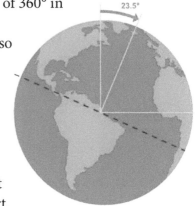

The Earth's Tilt [72]

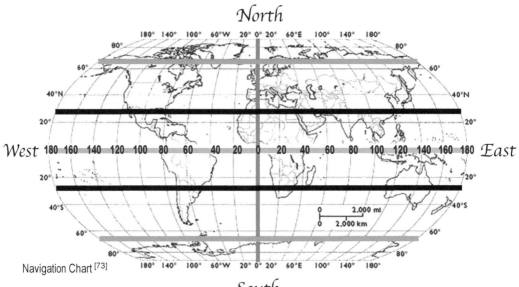

North

West *East*

2,000 mi
2,000 km

Navigation Chart [73]

South

I observed the similarities between this chart and the degree wheel. We have chopped off the North and South Poles so we can indicate degree numbers in accord with the grid lines. Although the earth is spherical shaped, it still qualifies as a complete circle of 360°—our magic number. There are light-colored grid lines in the background running East and West which we term as Parallels of Latitude and those running North and South we term as Meridians of Longitude. Accordingly:[67]

- ❖ The 0° Parallel of Latitude (orange) is named the Equator and divides the earth into its Northern and Southern Hemispheres. The 0° Meridian of Longitude (red) is named the Prime Meridian and divides the earth into its Eastern and Western Hemispheres.

- ❖ The black Parallel of Latitude line at about 23.5 degrees north of the Equator is the Tropic of Cancer representing the most northerly latitude at which the sun is directly overhead. (summer solstice). It happens around the 21st of June signifying the beginning of summer and the Northern Hemisphere's longest day of the year. Conversely, it is the beginning of winter and the shortest day of the year in the Southern Hemisphere (winter solstice).

- ❖ The black Parallel of Latitude line at 23.5 degrees south of the Equator is the Tropic of Capricorn and represents the most southerly latitude at which the sun is directly overhead. This happens around December 21st and marks the beginning of summer in the Southern Hemisphere and is its longest day in the year (summer solstice). Conversely, it is the beginning of winter and the shortest day of the year in the Northern Hemisphere (winter solstice).

- ❖ The green Parallel of Latitude about 66.5 degrees north of the Equator is named the Arctic Circle. The green Parallel of latitude located 66.5 degrees south of the Equator is named the Antarctic Circle.

- ❖ There are three major climate zones classified as Torrid, Temperate, and Frigid extending from the Equator north and south.

❖ Planet Earth rotates on its axis 15 degrees every 60 minutes. Since it is depicted as a complete circle, we determine the number of time zones by dividing 360/15=24. Accordingly, you will notice there there are 24 Meridians of Longitude designated at the top and bottom of the chart. Great Britain, established its Prime Meridian 0° in 1721 to be in Greenwich, England, and we have followed that designation on our charts.

Time Zones

Throughout the centuries, governments around the world arbitrarily have chosen their own standard for designating time and time zones. Generally, determining the time was derived by the reading of angles and shadows cast by the sun on a dial as the earth rotates. However, when charts came along, the zones had to be referenced to a longitude meridian designated as 0°. Many chose a meridian that crossed their own country, but the most popular, as mentioned previously, was that selected by Great Britain; and being our mother country, we embraced that terminology—Greenwich Mean Time (GMT)—as our reference point. Since time, distance, and speed are the essential variables in navigation, you continually will be making adjustments to your calculations based on where you happen to be geographically at a given time relative to GMT time.

Applying what we have learned in this session, Liverpool is five 15 degree time zones earlier than Salem; look at the chart and count them. Every 15 degrees—the distance between two meridians of longitude—changes time one hour. Therefore, the clock time in Liverpool is five hours ahead of Salem. If it is 1:00 p.m. (0100) in Salem, it is 6:00 p.m. (0600) in Liverpool. Obviously, you must adjust your schedule to be synchronized with your destination, or you will miss an appointment and likely miss a sale.

An additional application from this session is that we departed from a location with latitude and longitude coordinates of 42.5195° N, 70.8967° W; and are headed to a location with coordinates of 53.4084° N, 2.9916° W. I had to check the chart to confirm the accuracy of these readings. "Young man, you now have a better understanding of navigation charts. We will use them in every teaching session moving forward, and every day you spend on the big waters. I commend you for your careful attention and give honor to your parents for passing to you an unusually gifted mind. I predict nurture and nature will take you far!"

We had several more productive sessions in the coming weeks; however, with only a few more days before reaching Liverpool, some disturbing weather changes moved in on us. The sea was becoming increasingly turbulent and worsening by the minute. We constantly were adjusting the sails to combat the strong winds and swells. One canvas ripped like an old bed sheet as a beam split, fell, and gouged a large hole through the foredeck. I went below to inspect and found water in the hold but saw no holes in the hull. The water had that good smell to a grateful seaman.

I returned to Captain Chapman to report that there was no immediate alarm below. However, he had taken the wheel to relieve the seasick helmsman who was holding tight to the gunwale and heaving. The Captain's foreboding look concerned me. I remained close, helping different crews as they struggled with various emergencies. During a brief lull, he called me to his side and shared an aspect of this voyage that I had not heard. He revealed that, rather than risk unrest among the crew before the fact, he had withheld from us one of his fears about this particular voyage—a route he had taken previously. He continued by revealing that England and the British Isles are very vulnerable to storms, high winds, and turbulent seas. The English Channel and North Sea in particular, not only are treacherous; they are considered to be among the most dangerous waters in the world. Hundreds of shipwrecks and lives lost have been recorded throughout history. "The worst omen is upon us—a westerly wind it is!" came his despairing cry. He ordered that I inform the crew but not alarm them further if I could help it. The alarm already was more than intense without a word spoken. Nevertheless, I put on my best[68] seaman's face and moved about calmly and confidently.

> "...when it stinketh much, it is a sign that the water hath lain long in the hold of the ship; and on the contrary, when it is clear and sweet, it is a token that it comes freshly in from the sea. This stinking water therefore is always a welcome perfume to an old seaman; and he that stops his nose at it is laughed at, and held but a fresh-water man at best."[68]
>
> – William Boteler 1634

I made the rounds holding to whatever I could grab while wiping the spray from my face and eyes. The *Batavia* pounded and churned as the angry conditions held fast. I checked below again, and though there still was no water

coming through the hull, it was pouring through the opening in the deck at an alarming rate. We have to man our pumps immediately and assign two crew members to relieve each other. They are old, manually operated, and are very physically demanding. We found that only two were operable. After priming, we got them pumping while I reported to the Captain. Strangely, he was speechless and dazed but seemed to understand my message as he struggled with the resistant wheel. I returned below deck with another crewman, and both of us relieved the pumpers. We continued this process for hours unending; we rotated every 15 minutes as our arms tired quickly though our resolve was unyielding.

At my next break, I checked back with the Captain, and we agreed to lower and tie the remaining sails leaving us to drift at random; however, we were too late because only remnants remained. I returned to the pumping station to take my turn for another 15 minutes. We were in a standoff—not gaining but not losing. The storm raged on through the night and the next day…and the next… and the next.

A ray of hope brightened our spirits as the sun rose on the 8th day; we had drifted into the English Channel, and the west wind was pushing us toward the coast of Belgium. The pumping continued, but only three of us could continue. The resistance had overcome the will of the others; their arms were limp and paralyzed. I took over one pump and assigned the remaining two to the other.

In time—an eternity it seemed—the wind began to dissipate, and less water was flowing through the hole in the deck. We were able to pace ourselves better and take longer rest breaks. We had eaten very little during the continuing ordeal, and along with the continuous strenuous exercise, all had lost lots of weight. I had to run a piece of twine through two of my belt loops and cinch up the slack to keep my pants from falling.

On the 14th day, we stopped the pumping and fortunately found ourselves at the mouth of the Scheldt River off the coast of

Manual Bilge Pump [75]

Belgium. We hoisted the remains of our sails and crept to Antwerp about 55 miles upstream. When we entered their very bustling harbor, there was about an hour wait until we were given an open space to secure our ship. While waiting, Captain Chapman made some arrangements through the dock office to recommend a place for the crew to spend a night on the town at his expense. A famous local tavern was selected to provide all the services and especially a feast to be remembered. He promised to join in on the merriment and pay the bill after completing the docking matters.

All were dismissed with no curfew, and as the saying goes, they scattered like wharf rats. They were so happy to be safe and secure on shore again. He asked me to stay with him to help complete the docking procedures,permission requirement, and updating our ship's log.

He requested that I update the log book while he attended to the other matters in the dock office. He estimated it to take about an hour to complete after which the two of us would head to the tavern.

We separated, and I reported to his quarters but could not find the log in my brief initial look; but after a more comprehensive search in every nook and cranny, I actually found two log books, strangely covering the same period of time. I turned back quickly in both to the day of our departure from Salem and reviewed a few entries to familiarize myself with the manner and style Captain Chapman used. Mysteriously, I immediately noticed conflicting entries

The best laid plans of men and mice often go awry[69]

—*To a Mouse*
by Robert Burns

covering the same dates throughout the logs. The further I examined, the more obvious that one of the logs was counterfeit. I recalled my first conversation with Captain Chapman, and how I was somewhat troubled because of him withholding cargo information. He had brushed off my questions dismissively by saying we would have lots of money to buy whatever we wanted. My initial concern seemingly was being validated.

The plot really unraveled when I found an entry in the bogus log book describing a change in our itinerary from Liverpool to Gothenburg in order to escape a storm. Further, it portrayed an unspecified incident with land pirates while the crew was on leave and he remained alone on board. The official log did not mention a pirate incident nor claim about docking in Gothenburg. Clearly, we were 650 miles from Gothenburg, and no pirate incident had taken place. It was a fictitious story and an obvious intent to switch the logs at some point in time. The motivation was uncertain, but there were several troubling possibilities.

The Admiralty Laws, forbid the unauthorized changing or falsifying a ship's official log; it is a criminal offense. Furthermore, I had pledged my accountability to the ship owners when they endorsed my hiring. Within 15 minutes of being in the Captain's Quarters, my course of action had been decided for me without further thought. First, the apparent culprit needs to be given an opportunity to explain; so I decided the most diplomatic way was to be coy and make no accusations.

I hurriedly brought the official log up to date with the current entries and made a couple of minor corrections to give me grounds to open a discussion with Captain Chapman. I placed the bogus log unchanged back in the obscure location where I found it and made my way back on deck to wait for the Captain's return. I was hoping to avoid an unpleasant confrontation. I planned to not reveal everything I knew, but to carefully test with bits of information to see if I were treading on a sensitive subject. If so, the Captain surely would react accordingly. However, if he had a plausible explanation, there would be no cause for either of us to be troubled.

Soon he came from the dock office sharing that everything was in order there and asked if I had completed updating the logbook. I casually reported that the log was now current after making a few corrections. He immediately looked startled and questioned what I had changed in the logbook. I calmly stated that the corrections were computation errors and an incomplete chronometer entry. Then he asked if I changed anything about the stop in Gothenburg!

Without me raising the slightest hint of wrongdoing, Captain Chapman had made a crucial mistake with his question—an admission of guilt. It was a critical lapse in his memory because there was no mention of Gothenburg in the official log—only in the bogus log. Obviously, he did not realize that I had discovered the evidence he was hiding as well as some sort of cunning plan he had in mind.

I responded with a quizzical look and asked why would there be an entry about Gothenburg in the logbook when we were yet to arrive there. Stunned for a second, he exploded and became irrational and highly defensive. He accused me of being insubordinate and conspiring against him. Furthermore, he was suspending my leave and ordering me to remain on the ship until further notice. Also, he was leaving to join the crew at the tavern and would let me know on his return what official action he was going to take against me. He angrily stalked off without further comment as I stood convinced that my suspicions had been confirmed.

> Oh,
> what a tangled
> web we weave . . .
> when first
> we practice
> to deceive.[70]
>
> —Marmion
> by Sir Walter Scott

There must have been a dozen other ships coming and going within the harbor as I paced the dockside. Fate, providence, coincidence, or genuine pure luck prompted me to execute my plan as though it was meant to be. The good ship *Perfect* was preparing to depart for Boston within the hour—close enough! They were short-handed, and the Captain offered to trade me a ride home in exchange for my services. Within a few minutes, I brought my clothes in a box, boarded, and was shown my sleeping quarters along with a trunk for my belongings. My garment box was a little heavier now because of an extra item hidden beneath my clothes—the *Batavia's* bogus logbook...

I never saw or heard from Captain Chapman again. We did not depart on the best of terms; he likely would agree. In some ways, I feel remorse. He had gone out of his way to prepare me better to become a Captain. He was an expert teacher with superior knowledge of navigation and thoroughly demonstrated that he was highly skilled at it.[71] I owe him a lot—a favor.

However, we happened to disagree on a matter of integrity, and it is an odd coincidence that both of us are held accountable to the same employer. Upon my return to Salem, I presented to the ship owners evidence indicating that a potential criminal act was being crafted; however, my exposure of his plan prevented it from being executed. From that perspective, I returned him a favor. Had it continued to fruition, perhaps it would have cost Captain Chapman more than his job. The outcome for me remains to be seen; regardless, my conscience is clear, and I am at peace.

The Moving Finger writes; and, having writ,
Moves on: nor all thy Piety nor Wit
Shall lure it back to cancel half a Line,
Nor all thy Tears wash out a Word of it.[71]

—*Rubaiyat of Omar Khayyam* by Edward Fitzgerald

Belshazzar's Feast [76]

CHAPTER 7
Fiji

Moana Beach [77]

The outcome of the meeting with the owners of the *Batavia* was favorable on my behalf. They judged that I acted responsibly and honorably in representing their best interests in the matter of the ship's log. Consequently, I—still in my teens—was promoted to the dual assignment as Mate and Trading Officer on the *Clay* under the command of Captain Benjamin Vanderford. We currently are en route to Fiji to acquire sandalwood and beche-de-mer—known as sea cucumbers.

The latter is a marine animal that looks like a cucumber, and is found in abundance on the ocean floors of the South Pacific Islands.

Beche-de-mer [78]

After being harvested, they are processed into dry food products and sold at very high prices in Manilla and the Asian markets. Beche-de-mer is an edible delicacy but especially is in demand as an aphrodisiac. Medicinally, it is used as a treatment for arthritis and cancer.

The appeal for sandalwood preceded that of sea cucumbers for decades, but over-harvesting has severely depleted the supply. Consequently, the demand for the wood and especially its extracted oil, has increased along with the price.

Both the wood and the oil produce a distinctive coveted fragrance that has been highly valued for centuries. The wood is used widely in religious ceremonies and for aromatherapy. It is a primary ingredient in the most aromatic scents, cosmetics, expensive perfumes, incense, soaps, and for skin conditions including scars and wrinkles. Medicinally, it is lauded as a highly effective treatment for skin conditions such as acne, scars and wrinkles; and internally as an effective treatment for urinary infections and stomach disorders. Clearly, Captain Vanderford is not pursuing ordinary products in this venture. Beche-de-mer and sandalwood along with its oil are three of the most sought after items in the trading world.[72]

I am excited to be a part of this voyage and consider myself fortunate to be under the command of one of Salem's most famous shipmasters. Not only is he notable as a navigator, having circled the world more than once, he is also considered to be the best of all Salem traders. Since this is my first voyage as a trading officer, I am eager to learn this skill from the master and to continue refining my navigation skills. He knows quite a bit about Fiji, having already been there twice. In fact, he has the distinction of being on board the first American ship to have visited the Fiji Islands.

We were under smooth sailing conditions all morning, so he spent a lot of time detailing what he expected of me as the Mate. Further, he gave a general overview of what I could anticipate on arrival in the Islands. He especially had my ear when talking about our first major decision—the choosing of our route to Fiji. He contrasted the west passage around Cape Horn and the East passage around the Cape of Good Hope with advantages and disadvantages, but both portrayed as equally dangerous.

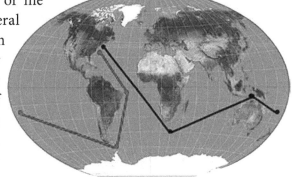

The East and West Routes from Salem to Fiji [79]

Cape Horn is at the southernmost tip of South America where the Southern and Atlantic Oceans meet the Pacific seemingly in a battle for supremacy. The

Shipwreck Off Southern Africa [80]

Cape of Good Hope is near the southern tip of Africa, although Cape Agulhas actually is the southernmost point. Regardless, the Indian and Atlantic Oceans converge in the area creating intense turbulence and very high waves similar to those found at Cape Horn. He warned that either choice can present the most terrifying experience a mariner can face. Both are reputed to be sailors' graveyards.[73]

"William, I do not want to frighten you unduly. I have rounded safely both of the Capes several times, but you should know what to expect in advance. I just doubt you ever have experienced waves 40 feet or more—bobbing and rolling your ship like a cork. Nevertheless, we will be prepared for whatever the conditions and soon we must decide which direction to take."

I asked him if one way normally was not better than the other. "William, the decision actually is like a comparison of Scylla and Charybdis.[74] Have you heard of them in Homer's stories?" I told him I only had heard about Homer from Professor Hacker; he once shared his tale about the Trojan Horse to illustrate the value of using the element of surprise; but I was interested in hearing about his analogy using those unusual names.

Odysseus tied to the mast and being lured by the sirens [81]

96

"Well William, Homer also wrote about Odysseus, a legendary character in that same Trojan War Professor Hacker referenced. He was a traveler and sailor of sorts—like us. After ten years he was trying to get back home in his ship when he encountered several mythological characters trying to discourage him. First, there were beautiful women known as sirens onshore. Their goal was to lure him with their hypnotic strains and cause his ship to crash on the rocks. Being a wily sea captain, he naturally outsmarted them by plugging the ears of his crew with beeswax and having them tie him to the mast. He wanted to hear their message revealing the ability to foretell the future, but he knew that knowledge would lead to evil advantages. Even though he begged to be untied contrary to his original command, the crew obeyed, and soon the hypnotic attraction faded in the distance.

"Once past the sirens, two additional mythological monsters appeared on each side of a stormy strait. Scylla was personified as the reef rocks and Charybdis, as a whirlpool. If the unfortunate Odysseus and crew came too

Scylla and Charybdis [82]

close to Scylla, they would crash, or if they came too close to Charybdis, they would be sucked down and drown. Hence, a saying was born to describe a dilemma in life where we have to choose between two equally undesirable options, i.e., we find ourselves 'between a rock and a hard place; or between the devil and the deep blue sea, etc.'

"So William, Scylla and Charybdis are personified in the two Capes—equally menacing. Nevertheless, I have made my decision. We will sail west around the Horn taking the Drake passage. The trade winds and ocean currents likely will be better this time of year. That route provides two additional alternative escape routes. We can take shortcuts by-passing the Horn entirely by crossing through the Strait of Magellan or the Beagle Channel. That is the choice, and we won't look back."

After that, the Captain walked with me to every duty station introducing each crew member and explaining what our relationship will be and the procedures we are to follow. Then, we went to the ship's hold to familiarize me with our storage capability. We were not carrying cargo in anticipation of loading to the hilt our premium purchases of beche-de-mer and sandalwood. As the new Trading Officer, I was asked by the Captain how I planned to keep up with the count of our per-item purchases, such as with the sandalwood, which would involve hundreds of small pieces. I responded that I never had an occasion where we were buying in numbers that large so I guess I would just keep count by writing on my slate. "What if there were several hundred being boxed or sacked faster than you can write or have room to write, he asked?" "Maybe I should use an abacus or trust the seller to be honest in recording the number; how would you handle the situation, Captain?"

"William, an abacus is too cumbersome and, as to trust, no responsible trader in the world—honest or dishonest—would fail to verify the accuracy of calculations in a transaction. It protects both parties from mistakes that can ruin relationships, reputations, and make paupers out of both. I require that you trust in yourself and count the items. I will show you my simple system which you will have with you at all times—your hands. I suppose the process was developed by cave dwellers thousands of years ago; regardless of when, it is very efficient and simple."

The Captain gave me a demonstration using a container filled with grains of rice. I used my thumb tip to count the digits individually—three per finger—for each grain he slid forward on a board. For practice, he would stop, talk, and try to distract me while stressing that I not move my thumb from holding my place where I stopped.

When we reached the 12th grain, he told me to make a fist indicating I had accumulated one unit then, begin back with number 1 on the other hand. If the purchase was greater than the combined 24, I was to move a coin from a supply in one pocket to an empty pocket. For the final accounting, each coin in the formerly empty pocket represents 24 items plus the number pending in the progression on the next count. Captain Vanderford said he had counted over five hundred items on many occasions using this simple

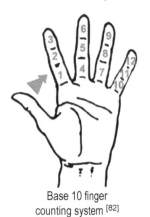

Base 10 finger
counting system [82]

system. I thanked him for teaching me this creative technique which I likely will find many applications in the future.

Now having completed our second week at sea, we have settled into a consistent routine. Fiji is some 12,300 nautical miles from Salem, so if we continue to average around 7 knots, the Captain estimates we will arrive in about 75 days (T=D/S). This includes a couple of port stops for food, water, and a short break on land. Rio de Janeiro, Brazil and Auckland, New Zealand are in our tentative plans provided there are no mishaps or foul weather.

Cooking on Sailing Ships [84]

In that regard, the ship's cook and I completed an inventory of our provisions and recorded a three month supply of salted beef, pork, fish, beans, biscuits, cheese, flour, limes, lemons, oatmeal, peas, rice, vinegar; and to drink: weak beer, rum, and water. Our meals were not like being at home but were satisfactory and sustaining. "Cookie" was very accommodating to adjust to our personal likes and dislikes and especially offer fresh fish that the off duty crew would catch trolling hand lines.

After I reported our inventory findings to the Captain, he seemed pleased and asked me to go with him on deck for some practice on a sextant. For background information, he gave me a thought-provoking summary of celestial navigation and how sextants are invaluable in helping find location.

"William, as a future Captain, your highest priority is to know your whereabouts at all times. Unlike onshore or within in sight of it, there are no landmarks at sea. When the fog lifted for you and your friend back in the Bay, in the story you told to Captain Chapman, you easily knew where you were from numerous familiar sights all around. Out here, you only see an endless expanse in every direction. Without knowledge, you and your crew can be hopelessly lost. I will see to it you never will find yourself in that predicament. However, you will not find it to be a simple task. My fellow captains agree with me that navigation is the greatest challenge of any trade. If it were simple, everyone could do it and reduce the demand and lucrative benefits. Therefore, it is a good thing that you will find it demanding. The moral to that analogy is that very few things in life that are easy prove to be exceptionally worthwhile.

"You have already proven to me that you could find our location on a

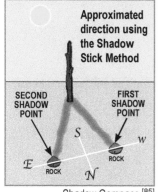

Shadow Compass [85]

navigation chart if you know our latitude and longitude. However, if they are not given to you and no one on board knows how to find them, it is up to you to save the crew and the ship. I will teach you how to do that.

"While it is true that you do not have landmarks to guide you at sea, you do have some objects almost as good—the celestial bodies.[75] The sun, moon, 58 stars, and four navigational planets, e.g., Venus, Mars, Jupiter, and Saturn, are more than sufficient to help us find our location. After darkness occurs tonight, we will focus on the stars, but for now, we have the sun beaming down in clear view; for

the sake of our eyes, we do not want to look at it directly.

"The ancients utilized the sun's shadow to create a makeshift device such as a compass and sundial. As you have learned to expect, a shadow cast by a stick driven in the ground moves 15° per hour in an eastward direction. Placing pebbles at the tip of the shadow every few minutes quickly creates an east to west imaginary line. Placing a stick at a 90° angle across the path of the pebbles will indicate a north to south line. Consequently, you then will know the direction of north, south, east, and west and the general course you need to travel.

Cross-Staff

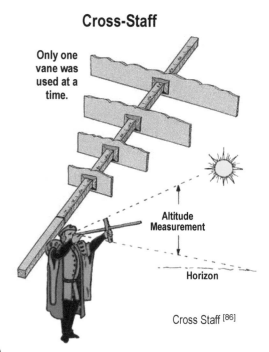

Cross Staff [86]

"In the illustration on land, we used angles made by shadows from the sun to give us the desired direction. At sea, there is no practical way or need to measure shadows. However, the early mariners learned to obtain all kinds of valuable information using crude devices to approximate the sun's distance from the horizon and other celestial bodies. By determining the degrees of various angles, the mariner can answer many kinds of time and distance questions.

"Fortunately William, you do not have to depend on approximations that can cause missing your destination by miles. You can be more precise by using refined instruments made for that purpose. I prefer a sextant which similarly—but more accurately—aligns the sun and horizon through a shaded eyepiece producing the exact angle between them and the observer. Once you know the degrees of the angle, you can determine that your latitude will be the same number of degrees

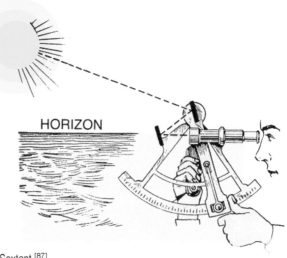

HORIZON

Sextant [87]

north of the equator. I want you to make and record a series of these sightings called "noon shots" starting at 15 minutes before and after noon to determine when the sun is at its highest. You are to do this every day near noon and immediately record the time of each shooting.

"Further, with a clear night expected, let's meet here after dark to learn an even simpler way to find our latitude."

Star light, star bright,

First star I see tonight,

I wish I may, I wish I might,

Have this wish I wish tonight.

—Anonymous

The Earth's Tilt [88]

Hours later and after a bowl of Cookie's stew, I found Captain Vanderford on deck pensively gazing upward at a pitch-black background dotted with countless millions of tiny bright lights.

In jest, I asked how many he could count using the hand and finger-digit system he taught me. "William, I can't tell you the exact number, but a professor friend of mine at Harvard College told me they outnumber all the grains of sand on all the beaches in the world. He said that astronomers had calculated there are about 10 billion galaxies—clusters of stars—in the observable universe and 100 billion stars in each galaxy.[76] It is not an exaggeration to say, but the universe is infinite, and the number of stars is uncountable.

Narrowing the scope, I can say with assurance there is one star and one constellation more important to mariners than all the rest, i.e., the North Star and the Southern Cross. I will tell you why.

"Following an imaginary axis through the center of the earth into space above the Northern Hemisphere, we find the North Star known as the Pole Star or Polaris. It is only 1° off to the side of the North Celestial Pole; therefore, its rotation is so small, it appears not to move—you will find it in the same relative position every night.

"William, I stop for you to grasp the importance of that fact; no other star is like it! It is a familiar landmark on shore that you can rely on to guide you to your destination. Of greatest significance is that a sextant shot determining the angle between Polaris and the earth's horizon produces a measurement that is almost identical to your current latitude.

"Take that one step farther. Knowing your latitude and with nothing else but a compass, you can find land at due east or west. This is known as 'dead reckoning.'

"Finding Polaris is relatively easy. Even as children, most of us learned to spot the Big and Little Dippers; but first we must get our sky ruler. With your arm extend-

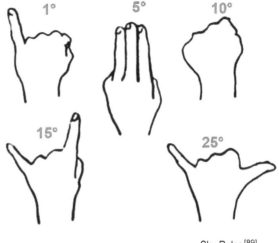

Sky Ruler [89]

102

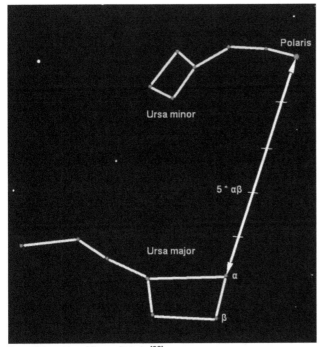

Ursa Major - Ursa Minor - Polaris [90]

ed, Polaris is about three fists on a line from Merak and Dubhe. Learn their names because they will become good friends."[77]

Convinced that I could determine our latitude from the Sun and Polaris as needed, I asked Captain Vanderford if he now would show me how to find our longitude. He said we are not finished with latitude. He explained that within a few days, we would be crossing the equator into the Southern Hemisphere where the North Star is no longer visible. Since we will be below the equator in the South Pacific for two or more years, I needed to learn about an alternative to Polaris. He stated that there is a relatively close equivalent known as the Southern Cross[78]—a part of the Constellation Crux. To be more precise, Acrux and Gacrux, aligned north to south create an imaginary line pointing directly to the Southern Celestial Pole. Using a sextant to measure the altitude between the Pole and the horizon will produce the number of degrees south of the equator the observer is located. Accordingly, that number will be the Southern latitude of the observer.

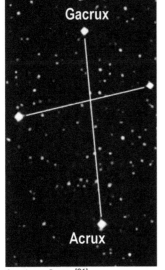

Southern Cross [91]

Captain Vanderford assigned me the responsibility of taking daily noon shots from the Sun and nightly from Polaris. I will do the same using the Southern Cross. He assured me that my ability to use a sextant and make these sightings would enable me and our crew to always know our latitude anywhere in the world; however, there was one final mountain for me to climb—longitude. It is a far greater challenge but equally necessary to determine our exact whereabouts at any time. He wants us to delay further navigation sessions for a couple of days so as not to interfere with the very special event just ahead.

The Crossing

I was not aware of anything forthcoming that was especially unusual, but he quickly schooled me about a 400-year-old tradition in the practice of the seafaring. It is the crossing of the equator where all the Shellbacks onboard plan a momentous day of initiation for all the Pollywogs—those crossing the equator for the first time. Since I am a so-called Pollywog and not keen on surprises, I am a bit uneasy over what this might entail. Nevertheless, I plan to follow the Captain's orders.

Captain Vanderford jokingly addressed the first-timers in preparation for tomorrow's big event. "Gentlemen, soon King Neptune and his Royal Court, along with the Queen, the Royal Baby, and Davy Jones will arrive onboard— the same as they have done since the beginning of sailing.[79] They will be joined

Crossing the Line 0° Latitude —The Equator [92]

by the trusty Shellbacks among our crew to watch a talent show. The performers will be all of you slimy Pollywogs hoping to become Shellbacks. You will have one hour to prepare for the entertainment, and I hasten to add, it better be good! If not, you will be issued a subpoena by Davy Jones to stand before the court for sentencing—a most likely development. In fact, throughout history there have been no pardons given or pleas of innocence accepted. Even if your talent pleases King Neptune, you can rest assured Davy Jones will not be amused.

"Davy lives in a Locker on the bottom of the sea watching over all the seamen drowned from shipwrecks. He is an irritable kind of fellow because of his depressing job. You do not want to upset him further because of bad acting, dancing or singing. Most of all, you need to make him and the King's Court laugh."

With tongue in cheek, we Pollywogs joined in the charade, but as predicted, our merriment and made-up acts failed to delight the audience, although there was quite a bit of laughing. In typical fashion and as expected, Davy Jones issued his subpoena ordering us to appear before the Court at the forthcoming crack of dawn immediately following breakfast—one which we found was too hot and spicy to eat.

Nonetheless, we were brought before a disguised Captain Vanderford, dressed as Neptunus Rex, to answer a variety of false charges and to be automatically sentenced. The punishment included wearing our clothes inside out, crawling through a wet grease-coated canvas tube filled with debris and slime.

The initiation concluded with each having to kiss the belly of the rather large Royal Baby who threw a smelly mustard-like concoction in our faces. Lastly, we took a bath in seawater and stood before the King to be crowned as Shellbacks. Each received a memento and certificate signifying to posterity that we indeed had earned a special status in the sailing world. It was a nice break from the seriousness of our regular routine and built closer camaraderie among all.

More Training

Back on the job, I took my noon sun shots with the sextant and determined we were approaching 15° South Latitude—my first entry south of the equator. The Captain had planned for our first Port of Call to be at Rio De Janeiro, Brazil, which our chart shows is at 22.90° South Latitude—about eight degrees south of our present location. Each degree of latitude is about 60 nautical miles, so we are about 466 nautical miles away. At about five knots per hour, we will travel about 120 nautical miles each 24-hour day. Therefore, we are about four days away from Rio. In our southward compass assisted course, we are staying in telescopic sight of the east coast of South America.

Captain Vanderford shared his strategy for stopping and remaining overnight in Rio. We have been at sea for 35 full days (24 hours each) and traveled over 4,000 nautical miles from Salem. He knew from previous experience that we needed to dock, get additional water, fruit, and spend some time onshore for the crew to experience a change of pace and routine—perhaps some fun and frivolity. He had learned that the simple act of recovering one's land legs away from the constant pitching, rolling, yawing, heaving, swaying, and surging is very therapeutic and renewing. He believed the downtime lost would be recovered quickly because of improved attitude and morale.

Further, he added, because we now are sailing south of the Equator, I needed to practice making some nighttime sightings of the Southern Cross to determine latitude rather than using the North Star which is no longer visible.

The transition to finding latitude in the Southern Hemisphere was quite easy. Once I accurately found the Southern Celestial Pole with a sextant and compared its altitude to the horizon, the reading in degrees is the same as our latitude south of the equator. I repeatedly demonstrated my ability to demonstrate this to an impressed Captain. With a grin and encouraging pat on my back, he said that we are ready to move forward, addressing my frequent request to learn how to find longitude.

The Chronometer[80]

"William, prompted by thousands of lives lost from shipwrecks due to faulty longitude assessments, the British Government passed and funded in 1714, The Longitude Act to reward anyone who could find a practical and precise method for determining a ship's longitude. A fortune awaited the inventors and John Harrison, an English carpenter and clockmaker, is credited as the major contributor to this accomplishment with his Marine Chronometer. His device simplified what historically had been the most difficult challenge in all of navigation.

"The better you understand the problem, the more you can appreciate the solution. Even if the early mariners were sailing at constant latitude determined by a fix on the noonday sun or the evening North Star, they had no reference point north to south because of the earth's continuing rotation. Therefore, navigators came up with many different makeshift ways of estimating and approximating longitude—all fairly complicated and imprecise. Astronomers believed the only solution possible would be found

The Scilly Naval Disaster of 1707— prompted The Longitude Act [93]

in the sky by sighting and mapping the heavenly bodies. Galileo came closest by devising an intricate process using the orbits of planet Jupiter's satellites, but it would require having an accomplished astronomer on board to execute it. This and other endless futile efforts proved impractical and too difficult.

"Centuries before in a different domain, an ingenious fellow, William of Ockham, a Franciscan Friar, came up with a notion that likely influenced John Harrison's reasoning in his pursuit of a solution. It was a problem-solving method that not only offered a parallel to finding an answer to the longitude quandary but an approach for people in general to use when making everyday decisions involving a variety of situations. His problem-solving principle became known as 'Occam's Razor.'[81] It has nothing literally to do with shaving other than as a metaphor

John Harrison Chronometer [94]

for sharpness. He logically concluded that in decision making, where there are several possible choices, choosing the simplest approach with the fewest assumptions and variables will always produce the best results. It sounds so elementary that it is a wonder why it works so well. However, historical efforts to disentangle the longitude predicament were confounding due to the complexity of the remedies.

"John Harrison applied Occam's Razor—unwittingly or ingeniously—to perfection. He reduced the troubling variables to the minimum. He chose to use "time" rather than celestial bodies as having the fewest variables. Further, he so simplified his clock that it remained accurate within seconds and was unaffected by temperature, weather, and the motions of a ship at sea. The gears were made of a special wood unaffected by the corrosive ocean environments and required no lubrication. William, the complex problem that eluded the best minds for centuries was solved, not by a seaman, but by a carpenter and watchmaker. Strangely, there is no record of him ever having been on board a ship. Thankfully, we now will take advantage of his invention.

"After finding with our sextant, the current local noon (high noon) sun angle compared to the horizon and recording the local time, we need only one more variable to know our exact longitude. Our reference point is 0° meridian located at Greenwich, England and our chronometer always is set at Greenwich local time. Notice that we are about to compare two different locations based on their respective local times. We currently are west of Greenwich off the South American coast of Brazil. The sun appears to rise in the east because the earth's rotation and moves westward one meridian each hour. Therefore, we must subtract an hour from Greenwich Mean Time (GMT) for each local-time hour to locate our exact longitude. We interpolate (estimate) for points not falling exactly on a meridian. For example, your recent rounded noon sighting indicated we were at 8.05° S latitude at noon Brazilian time which is 3:00 p.m. in Greenwich, according to our chronometer—three hours difference. This places us close to 34.9° W. of Greenwich which our world map shows is just off the coast of Recife, Brazil. Checking for accuracy in my British navigation almanac, I find Recife's official location listed as 8.0522° S, 34.9286° W. It is close enough, William, and simple enough thanks to John Harrison."

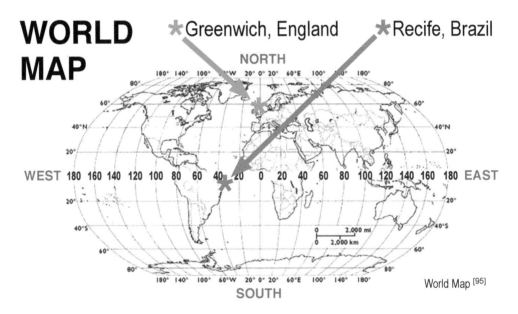

World Map [95]

I continued with my daily routine overseeing each duty station and focusing on sextant noon shots of the sun and at night, the Southern Cross. For further practice and my quest for knowledge, I began memorizing the names and locations of all 58 stars and four navigational planets which along with the sun and moon should prevent me from ever being lost at sea or on land. Captain Vanderford was pleasantly stunned to learn that to be my goal. He said that most navigators only learn three or four, so if I reached that level, I might be the only navigator in history to have done so. He said it would be a very impressive achievement to include on my dossier when applying for Captain status. As I contemplated that possibility, my thoughts were pleasantly interrupted by the lookout's cry of "Land Ho." We were entering the harbor of one of the world's major seaports.

Rio De Janeiro, Brazil — 22.9068° S, 43.1729° W

We will be in Rio overnight primarily for restocking water and fruit. As well, shore leave will be provided those who wish, but all must be on board to depart by noon tomorrow.

Regardless, immediately after mooring, the Captain, Cookie, Joko (the ship's store crewman) and I proceeded to a supply center to make our planned purchases. They had a bilingual speaker so our shopping went smoothly and included delivering our wares back to the ship's hold. With our mission completed, shore leave was

View of the Rio Bay as seen from the convent of Santo Antônio [96]

declared by the Captain. Some preferred to stay on board to catch up with much-needed sleep. However, most wanted to go into the city for a night on the town and to mix and mingle with people and culture of one of the world's largest and most unusual cities.

My preliminary research revealed that the city was founded around 1565[82] by explorers from France and Portugal, but the Portuguese ultimately prevailed as did their language. Although there are slight similarities, my Spanish fluency would not be very helpful. I learned enough phrases to assist with basic communication. My group of four decided to hire an English speaking guide to familiarize us with the territory. His father was an American international trader in the sugar business; his mother was Brazilian and a teacher.

We walked throughout the city and visited several interesting sites as he shared the history and culture. We learned that coffee, sugar cane, gold, diamonds, dyewood, rosewood, and classical guitars are prominent products, and music is the national pastime. The rosewood particularly is very rare and expensive because of its use in building the world's finest classical guitars.

Rio is the largest seaport in the Americas for receiving slaves. Well over a million have been imported from Africa in the last century primarily to harvest sugar cane. Sadly, the practice continues to flourish at this time in spite of the country's strong religious influence. Roman Catholicism not only is the main religion in Brazil, but it has more practicing Catholics than any country in the world.

Our guide took us to a café-like eatery for our night meal, which featured lots of tasty chicken and beef dishes—barbecued and spicy—with trimmings and incredible desserts, unlike anything we had ever eaten.

We were lucky, according to our guide, that we are here when the Carnival

Carnival [97]

Festival[83] is taking place. It dates back to early 1723 and is celebrated in honor of the gods and the great waters. The event features parades, music, dancing—all in masquerades.

We had a busy fun-filled day and evening but thought it best to return to our ship to rest before departure.

En Route to Cape Horn

The following 22 days were relatively routine and uneventful until we reached the Cape. We were approaching the place known as the sailors' graveyard where thousands have died, and countless ships went down. I must admit to some anxiety until the Captain spoke and settled our nerves.

"Gentlemen, I will update you about our plans for the coming week. In a day or so, we will be approaching Patagonia, a region on the southern tip of South America shared by Chile and Argentina. To safely reach the Pacific side of Patagonia, we either will use Magellan's Strait, the Beagle Channel, or Drake Passage.[84]

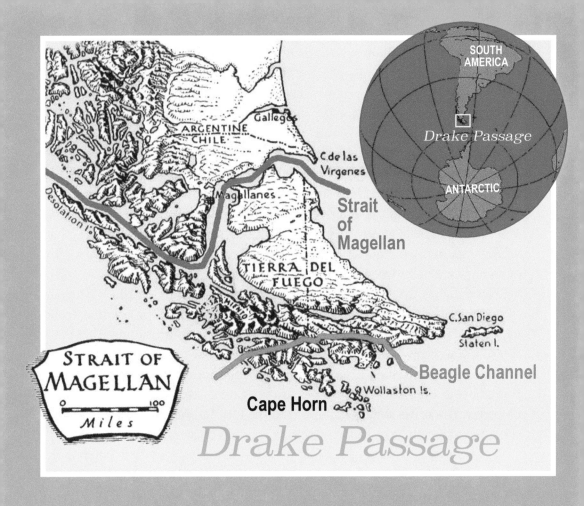

Strait of Magellan and Beagle Channel [98]

"As you are aware, the flag signals we have been seeing continue to warn us of extremely treacherous ocean turbulence and such strong winds that Drake Passage just south of Cape Horn is virtually impassable at this time. One of the vessels reported waves approaching 90 feet at their crest. We will hold up just inside the Strait and wait for the weather to moderate before making our decision on the route we will take."

On schedule, we arrived and anchored near the north bank leeward from the wind in a calm, secure area. It was peacefully relaxing while all were served one of Cookie's special dishes. It was a chopped beef and gravy concoction served over hardtack biscuits along with a pint of beer to help drown the weevils often present in the flour. Cookie overlapped the biscuit halves somewhat like house shingles, then layered the chopped beef and poured the salted and peppered gravy over the pile for a meal fit for a king. He even had a crude name for his shingle dish that he said had been handed down for hundreds of years. We ate on deck by the light of a star-filled sky, and when finished, the Captain said he wanted to talk to us about several matters. It is infrequent to have all shipmates together at one time because at sea we have around the clock staggered watches and duties. However, this was a special opportunity to hear from this man everyone respected so much for his character, intelligence, and seamanship. We especially loved to hear his sea adventure stories and hoped we could persuade him to tell us some. We listened intently.

"Gentlemen, since I am the only person on board ever to have been this far down into the Southern Hemisphere, I am offering to share some things that you might find curiously interesting. It has to do with cannibals and giants. This is not a fireside ghost story but an account of some strange events and developments in the history of this territory. They may be factual, mythological, or somewhere in between.

"Nevertheless, in the early 1500s, Ferdinand Magellan,[85]—the Portuguese explorer—sailed under the flag of Spain through this very strait with a fleet of five ships. His goal was to find a western route to the Spice Islands. There were 270 combined crew members, but only 18 made it back to Spain; Magellan was not one of them, having been killed in the Philippines. Although he is credited as being the first person to circumnavigate the world, this technically is incorrect. It was his ship that completed the feat, but many believe his trusted servant and beneficiary, Enrique, actually should be awarded that honor.

"Regardless, a strange thing happened when Magellan first entered this strait, no doubt trying to find a shortcut to avoid rounding the Horn. Antonio Pigafetta, his chronicler, wrote an account of them seeing natives that were twice their normal height. Some were up to 15 feet tall—they were giants. Magellan sent one of his men to engage with the giant and found him to be friendly. He brought the giant to confer with Magellan and several other eyewitnesses who only came up to the giant's waist. The giant pointed to the sky obviously thinking they had come from above—perhaps gods. Magellan named him Patagon from the Portuguese word meaning 'Big Foot' and the territory he named Patagonia meaning 'Land of the Big Feet.'

"Fifty years later, Sir Frances Drake's chaplain, Anthony Knivet, reported that he had seen dead bodies in Patagonia which were 12 feet long. The same year, Englishman William Adams reported a skirmish in the area between a Dutch ship crew and unnaturally tall natives. Accordingly, some of the early maps labeled the area as 'the region of the giants.' A 1776 report from crew members of *HMS Dolphin* described sighting natives in Patagonia who were up to 9 feet tall. Stories of sightings of Patagonian Giants have circulated all over the world for a hundred or more years.

"I am telling you this story because of geographical relevance. Our ship is anchored at this moment in the heart of Patagonia. Regarding cannibals, our destination is to Fiji, which is not named the Cannibal Isles as a compliment. That is another story for another time."

The Captain's story about "giants" touched a curious nerve in the entire crew and prompted several probing questions. The first crewman asked for the Captain's explanation of those reported sightings and if he had ever seen a so-called giant.

His measured response is evidence of his wisdom. "I am slow to believe anything that I have not seen; yet, I believe in the existence of more things I have not seen than things I have seen. For example, I never have seen or felt the movement of the earth rotating, but I have seen evidence that the earth does rotate at great speed. To your question, I never have seen a giant as described, but I have seen paintings of giants and have spoken with people who claim to have seen them. I have seen skeletons of the deceased, which look as though they were exceptionally tall and giant-like. As well, I have read about the existence of giants throughout the Bible. Still, I have not seen a giant other than Cookie who is 6 feet and 7 inches tall—a genetic anomaly.

"However, men, you are about to spend the night in the land of the giants. If any of you happen to see one, please wake me so I can see it as well. Now, that is

Patagonian Giants [99]

your bedtime story for the evening. Have you noticed the change in the sky and wind direction? We could be in luck to make a move at daybreak, so get your rest to be ready."

Fair Winds and Following Sea
Captain Vanderford called us to muster at the sound of the Boatswain's whistle. He explained that we have a window of opportunity to move out. He wants to avoid continuing through the Strait of Magellan, although thankful that its mouth has provided us a tempo-

> "IF THE HIGHEST AIM OF A CAPTAIN WERE TO PRESERVE HIS SHIP, HE WOULD KEEP IT IN PORT FOREVER"[87]
>
> —THOMAS AQUINAS

rary refuge. The narrowness at certain places along with the twists and turns leave little margin for error if the wind gusts; plus, it is over 300 nautical miles through it. He has decided we will exit and continue southward to Drakes Passage then on to Fiji.

How thankful we are, Captain Bligh of the *HMS Bounty* waited for a month for the weather to calm. He finally gave up and sailed east back across the Atlantic to the Cape of Good Hope in order to reach Tahiti. At Cape Horn, the usual large waves and currents can be expected as the Atlantic, Pacific, and Southern Oceans converge, but with the wind being somewhat moderate and the sky not foreboding, Captain Vanderford feels the extreme turbulence will last only for a mile or so. We will reef the mainsail and furl the jib (roll and wrap) as a precaution at that time.

We have set sail and are about 200 nautical miles and two days before the westward turn at the Horn; therefore, we will need to hold up at sundown near the leeward tip of of Tierra Del Fuego to avoid crossing the point in darkness. I cannot cease reflecting that I soon will have sailed on my third ocean, visited or bordered five of the world's seven continents, and dozens of countries—still as a young man. I am grateful and feel very fortunate.

The hours passed routinely, and our reward was a quiet night anchored off the leeward tip of Patagonia. By mid-day, we were underway and could see the different colors of water at the confluence of the Atlantic and Pacific. By the time we were about even with the Horn, the wind picked-up, and the water became quite choppy. We reefed and furled our mainsail and jib; and passed through the danger zone with little resistance. Nevertheless, with Antarctica at our backs and South America in our faces, Captain Vanderford stepped to the starboard rail and respectfully placed his hat over his heart and kneeled

Waters of different density. [100]

in memory of the many thousands below in their watery grave. Without a word spoken, the entire crew followed the Captain's example. It was a chilling experience and somewhat eerie as we seemed to hear voices of our sunken comrades whispering, "Godspeed Brothers."

In time, we cleared the Horn and the cross-currents relaxed to let us go our merry way. We are sailing onward to Fiji another 2,000 nautical miles away hoping, despite a westerly headwind, to average five knots per hour, this should advance us about 120 nautical miles per full day and reach our destination in about two weeks.

The Captain also held frequent information sessions with the crew beginning with our most formidable threat—cannibals. He explained for that is the reason our destination is one of the few places in the sailing world that the insurance companies will not sell coverage for ships.

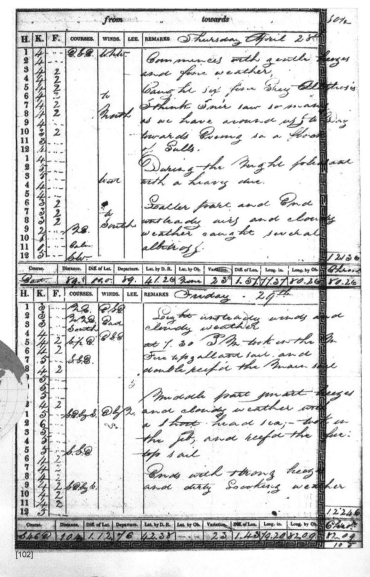

75° WEST 41° SOUTH

The *Charles Doggett* ship log entry on April 28, 1831, gives their location in the Pacific after they rounded Cape Horn

Cape Horn, *Charles Doggett* log, location 75° West 41° South[103]

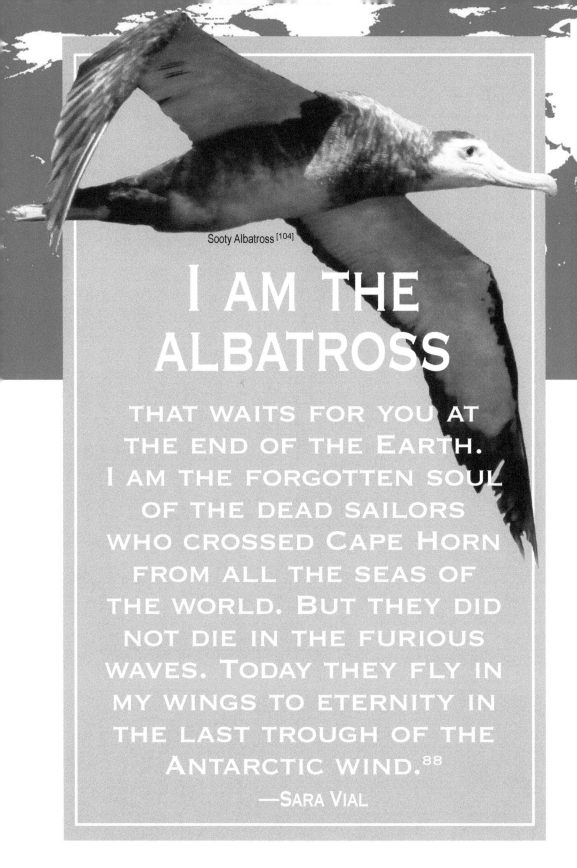

Sooty Albatross [104]

I AM THE ALBATROSS

THAT WAITS FOR YOU AT THE END OF THE EARTH. I AM THE FORGOTTEN SOUL OF THE DEAD SAILORS WHO CROSSED CAPE HORN FROM ALL THE SEAS OF THE WORLD. BUT THEY DID NOT DIE IN THE FURIOUS WAVES. TODAY THEY FLY IN MY WINGS TO ETERNITY IN THE LAST TROUGH OF THE ANTARCTIC WIND.[88]

—SARA VIAL

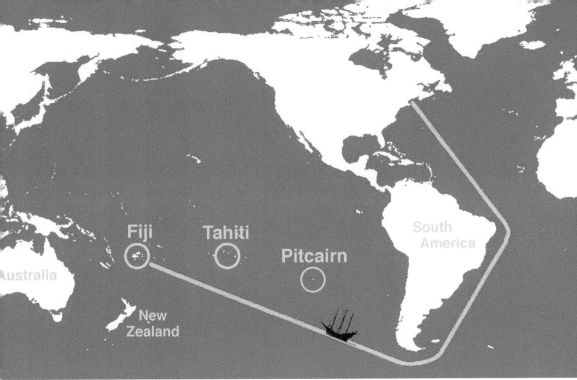

Pacific Basin Islands [105]

"Crew, the idea of eating human flesh is virtually unthinkable in most of the civilized world except in rare isolated instances where starvation and impending death presented it as the only alternative. However, encountering cannibals is an ever-present possibility in certain Fiji Islands which we naturally will avoid. I had a frighteningly close call with them early in my career when captured, but later released because I was too slender to make a good meal.

"Lurking somewhere in the islands is Chief Ratu Udre Udre,[89] who is believed to have eaten over 800 humans, and he is still alive at this time. There is no food shortage on the islands, so it is assumed, well beyond religious symbolism, that the chief has cultivated a taste for human flesh.

"Therefore, we must be prepared for the possibility of conflict with them and have a strategy to combat a potential attack. In jest, I suggest you stay thin; as well, the adversaries seem to welcome and embrace the foreigners who can speak some words in their native Fijian or Hindi tongue. I have language books in our resource room that will teach you basic expressions which I encourage you to learn.

"Also, there is strength provided by our numbers and weapons which they do not have. There are no ports as such, so we will operate from our ship anchored beyond the reefs from the various villages. Our utility boat will serve to transport us to and fro. As well, in most of the villages, I have established a friendly relationship with their chiefs. I bring them gifts and have become fluent in their language. They know why we are there and on arrival paddled

Cannibal Feast [106]

their dugouts to our shipside loaded with beche-de-mer. They are in high demand in the Philippines and China as aphrodisiacs and for use in medicines and soups. The natives harvest them by hand from the shallow warm waters surrounding the 300 islands.

It is the outer body—cut, cooked, smoked and dried—that is of greatest value. Learning how to prepare and cure them is a closely guarded secret which we need to learn. For now, we have to purchase them ready for the hold. The value of these sea cucumbers is so enormous that a shipload can bring a small fortune for the owners."

After several weeks in Fiji, we had accumulated 600 piculs, which is approximately 133 pounds per picul. A picul is the standard measurement adopted in the islands to be the amount one can carry comfortably on a shoulder pole. Along with several yards of sandalwood, we sailed to Manila with almost 80,000 pounds of beche-de-mer which we sold quickly for over $25,000. To put that amount in perspective, the average yearly wage back home for professional workers such as doctors and teachers is $300 per year and $200 yearly for laborers. One easily can understand with such stakes in play why the lure of the sea seeking in-demand products is so compelling. A single voyage can provide a life of leisure in early retirement.

Periodically, Captain Vanderford would meet with an international agent to exchange goods and money. This agent had connections to an Australian bank with an affiliation with a bank in Boston. I remained on board in charge of the *Clay*, and the remainder of the crew was given shore leave as a reward for our success. My napping in the Captain's quarters was suddenly interrupted by two grizzly characters with flintlock pistols drawn. I concluded that they were pirates with robbery in mind. I had planned a strategy if this sort of thing were ever to happen, so I went on the offensive immediately.

Island girl gathering tea leaves [107]

"Are you the mates your Captain sent to talk about the big-money job I have for you?" With a haughty laugh, the larger of the two responded in broken English, "We are here to get big money for certain but not a job; and no, our captain did not send us. It's a long story involving mutiny, lowly cannibals, and respectable pirates. Our Captain and all his men except the two of us had some unfortunate luck while harvesting beche-de-mer. The cannibals captured them, destroyed our brig, and …well it is sad to say, they are no longer with us. Fortunately, we escaped—with much of the loot—and chased off the savages; now, on behalf of the living and the dead, we are here to collect money to help us get re-established in our business. We are visiting all the brigs in the harbor seeking donations. We are confident you will want to cooperate."

"I am sorry to hear of your bad luck, and that you are too late for me to help your cause; our Captain already has deposited all our resources in an international bank. However, I can offer you some good news far greater than the few silver pieces I have in my pocket; and more importantly, make you very wealthy in the process. Please have a seat, light up a Cuban Cigar, and sip on a tot of rare Jamaican Rum, and let's talk about it."

The pirates put away their pistols and with a somewhat confused look, took a seat to enjoy the amenities offered while I explained my scheme.

"This is the deal, mates: First of all, you mentioned harvesting beche-de-mer when the cannibals confronted your group. Tell me what you know about curing them."

"Well, it is a very complicated process, but we were taught by the best in the islands how to prepare them; you can say that we have become experts. Anyone can scoop them off the bottom, but very few know how to cure them properly. First, the cucumbers must be kept wet and out of direct sunlight. Next, they have to be gutted and boiled in a try-pot like whalers use to render oil from blubber. Then they have to be rinsed, smoked, and dried in the sun or in batter houses. Generally, that is how it is done, but for the right kind of money, we can teach you how to do it."

"It is clear you have the knowledge and skill, but you do not have a ship or the know-how to take you thousands of miles to markets and savvy traders who will pay high prices for them. If we become business partners, it will only take a few transactions to set you up for life—as you said, the right kind of money; you can retire from pirating, and it is much more honorable. Your job will be to recruit the natives, collect and cure the cucumbers, and we will do the rest. We will set up regular buying trips and pay you a commission for each transaction you conduct with the natives. You will invest nothing but your obvious knowledge and can begin immediately building your fortune.

"The owners of our ship are The Rogers Brothers out of Salem, Massachusetts. They own several other ships including the *Quill*, to which I am scheduled to transfer on their voyage to Fiji as we are returning to Salem. You can test our relationship as soon as the *Quill* arrives in Fiji. Try it just once, and you will see how wealthy it can make you." After their second tot of rum and in a somewhat giddy state, we became business partners with an agreement to rendezvous in the coming months to execute our first trade.

Captain Vanderford and the crew returned by nightfall, and at the crack of dawn, we were sailing eastward to rendezvous with the *Quill*. As planned, I will become their Mate and return to expand our beche-de-mer business. With 300 islands in the archipelago, there will be more than enough territory to prospect for the *Quill* crew, me, as well for my new partners.

We arrived at our destination in Murders' Bay, New Zealand, where we found the *Quill* anchored and waiting for me. I transferred my sea trunk, and soon we were on our way back to Fiji under the command of Captain Kinsman out of Salem.

Fiji Club Dance[108]

Return to Fiji

The *Quill*, being a ship in the Rogers Brothers fleet, fortunately, has enabled me to extend my employment for another two years without having to return to Salem. We have discovered a figurative gold mine throughout the Fijian Islands because of the abundance of beche-de-mer in their surrounding shallow waters. Captain Kinsman generally was uninformed about these valued sea cucumbers and relied on me to explain all aspects of the finding, harvesting, curing, and marketing process. He particularly was impressed when he heard that our last transaction produced a $25,000 profit only as an add-on venture to our original mission.

I explained there was a well-established network of suppliers because of the friendly relationships we had built with numerous island chiefs. Every time our utility boat is seen approaching a landing, a celebration occurs with the islanders and chief waiting to embrace us with flowers and gifts. Naturally, the chief expects his customary gift. If there is a really big trade needing his influence, a muzzle load rifle as the gift can make it happen; a gun is worth a king's ransom to a chief.

Since Captain Kinsman never had visited a Fijian Island and was totally unfamiliar with their language, he wanted to go ashore with me as an observer of my trading style and methodology. Also, he was eager to get a better understanding of the people and their culture. He asked me for a general orientation prior to us leaving the *Quill*. I welcomed this as a good opportunity to blend into the conversation the most compelling quandary I have ever faced.

Fijian Couple[109]

"Captain, perhaps you will experience the same culture shock as I did when first setting foot on one of the islands—not the islands where the missionaries were beginning to become established but on those left unrefined. I came from puritanical roots where modesty in manner and dress were strongly emphasized and practiced by the ladies. Conversely, in the islands, you rarely will see a woman covered from the waist up and many, not at all. Intimate relationships outside marriage are accepted as the norm; and even then, the practice of having multiple partners and marriages not only is tolerated, it is encouraged.

"The early Christian missionaries are beginning to introduce reforms changing some of the practices to be more conservative, but their presence is sparse at this time. Do not be alarmed by the aggressiveness of certain female islanders in a sexual context. They will show no inhibitions and expect the same from you. During the previous two years, some relationships between our crewmen and certain Fijian women grew beyond casual affairs to marriage. The ceremonies were brief and conducted by the chiefs with no cost other than the dowry paid to the groom by the girl's father. The ultimate cultural adjustment to our conventional way of thinking usually comes with the introduction of the other husbands or boyfriends attending the wedding ceremony.

"So, Captain Kinsman, as I learned from reading Alexander Pope and other free thinkers, there are stages through which all of us are able to evolve in becoming more accepting or at least tolerant of customs that are dissimilar to our own. Even in matters we initially find to be outlandishly objectionable; frequent association, and familiarity in the context of prevailing practice, tends to make us more moderate.[90] In other words, it depends a lot on the environment and how it affects us personally—situational ethics. Pope wrote a poem about the subject—the three steps that can transform hate to love. Unfortunately, the principle of association can work both ways. If our choices are not wholesome, familiarity can lead us to embrace outcomes which are not redeeming.

"The preceding prompts me to share with you that I found myself in that dilemma. I befriended an island chief, Asham, and in appreciation, he actually gave me his daughter, Talei—a customary practice and way of showing gratitude in his domain. To refuse such a gift would have been an insult to the chief with far-reaching negative implications. He placed her hand, and a lock of her hair in mine and with eyes to the heavens gave a short incantation. I soon learned that by tribal tradition, the ceremony meant that Talei had become my wife. Although stunned, I was comforted to learn that she had no other husbands at that time to further complicate the situation."

As a dowry, the chief gave us a hut which became my home base for the remainder of time I represented Captain Vanderford as a trader throughout the Fijian Islands. From being total strangers and crossing paths only by circumstance and fate, we grew in affection, compatibility, and love. Talei's name translates in English to the word "precious" and none has ever been more fitting. Her beauty inwardly and outwardly indeed has been a precious event and blessing in my life. We spent a lot of time together during the *Clay* voyage, and with your permission, we hope to continue during this one. However, looking forward, there is an impasse. The island laws do not permit her to leave, and it is an unrealistic and impractical consideration for me to remain here permanently.

These were happy days and the time passed quickly for the

Fiji female dress codes after the first missionaries arrived [110]

crew of the *Quill* and especially for Talei and me. Captain Kinsman had been given a thorough orientation and introduction to every chief with whom we had built relationships. He will be quite capable of representing our owners in future ventures to the Fijian Islands.

We made two highly profitable trading jaunts to Manila and while there met up with my pirate business associates, established as a result of the thwarted robbery attempt. Amazingly, they had followed through on my proposal and had created a network of suppliers in new islands heretofore unexplored. They facilitated a third trade that was a financial bonus for our owners and indirectly for us. As well, the pirates were well-rewarded with income far greater than any of their stealing escapades. We made plans to continue using their services with the other ships in our fleet out of Salem.

We had a strange conversation before we departed for Fiji. The former pirates shared a life-changing experience that ironically was initiated indirectly as a result of their robbery attempt. While prospecting for suppliers of beche-de-mer, they had encountered a Methodist missionary in Manila who had been in Tonga considering establishing a mission. He befriended them with information that was valuable in their finding more suppliers to grow their business. More significantly, according to them, they were introduced to the Christian religion and its simple gospel of forgiveness, salvation, and eternal life.

Neither had ever been inside a church, met a minister, or heard a moral message, but long story shortened, they experienced a conversion and committed to a different path in life. Both had felt heavy guilt and remorse because of their lives as thieves and even admitted to having committed murder. The point of their opening their hearts to me was they wanted to offer sincere appreciation. Inadvertently, I guess I had helped redirect their path to a mission and purpose life of helping people rather than harming them. Tears flowed as we hugged and bid farewell perhaps never to see each other again.

Our assignment in Fiji is finished, and soon, we will be underway on our return voyage to Salem. Talei and I had separated previously in several circumstances; however, I was especially had mixed feelings and saddened this time to be leaving her in the wake of our parting ship. I promised to return.

Under Conviction

I remember distinctly January 5, 1823. I had been unusually troubled for several days as we sped westward on a friendly sea; I was not at peace although I seemed to have the world on a string, still yet to reach my 21st birthday. Perhaps it was being separated from Talei and the unusual nature of our relationship, or maybe it was the uncertainty as to my next venture after arriving back in Salem. I was in a state of confusion and discontent with no immediate explanation. I could not clear my head of the conversion experience my pirate friends shared. They were as low down as humans could fall, yet mysteriously became transformed into new creatures.

I woke from my restless sleep with a revelation as clear as the handwriting on the wall. My trouble was not in my head but in my heart and soul. I reflected back on my childhood experiences that had turned me away from religion and Christianity. The setting and presentation of the doctrine and principles were as morbid and unpleasant as the endless sermons I was forced to hear shouted from the raised pulpits. I daydreamed about the sea for hours just to survive the torture.

More so, I routinely observed behavior throughout the weekdays that contradicted the saintly fronts being displayed on Sunday—hypocrisy. I learned all about the seven deadly sins, e.g., envy, gluttony, greed, laziness, lust, pride, and wrath—not in the harbors and shipyards—but in the church houses. I rejected the pleas and righteous examples of my two brothers, who were Baptist ministers. I had no time or place for religion in my life, yet I yearned for the same contentment the pirates found. Perhaps I should have listened to my brothers.

In desperation, I borrowed a Bible and seemed led to numerous verses I had heard and were explained many times as a child. I came to realize that I was under conviction for doing the very things for which I was criticizing the church and others of doing. I was a sinner brought to my knees, confessing that it is not Christian teachings that fail to blend with a seaman's lifestyle. It has been some of my lifestyle choices that are incompatible with the principles of Christianity. No medicine or act of desperation could cure my hopelessness—only the supreme sacrifice of a Savior. I arrived back in Salem as a new man—redeemed, restored and saved by grace. The weight of guilt has been lifted. I can hardly wait to report to my parents and brothers.

CHAPTER **8**

William Driver becomes Captain

My brother, Joseph, was the first of my family to hear of my conversion and surrender to God's will. Joseph and brother Stephen have led me to the cross that I stubbornly resisted, but praise to God they persisted. Joseph said he would send the good news to our special friend, John Putnam, in Boston. He, also, has prayed without ceasing for my salvation.

To this point in my life, I have lived under "God's Permissive Will," making value judgments in accord with my selfish interests. Words cannot express what a glorious relief it is that I no longer have to think twice. Now I am under "God's Directive Will" comforted by the assurance that spirit-filled choices always are infallible.[92]

Strangely, I was ambivalent about being home again, visiting with family and friends. Much has changed during my four years in the South Pacific. The reunions were pleasant and refreshing, yet the lure of the sea is so powerful I soon became impatient.

Captain William Driver in Dress Uniform [112]

However, my restlessness dissipated in a marvelous way when I was summoned to meet with my most recent employers, the Rogers Brothers; they were owners of a fleet of Salem ships, including the *Clay* and the *Quill*.

I met with them in their harbor office with Mr. Nathaniel Rogers speaking on behalf of his brothers. "William, my brothers and I have discussed at length a proposition entirely contrary to our standard business operating procedures. We prefer our shipmasters to be at least 30 years of age with many years of sailing experience. In addition to the enormous investment in a ship and the advanced knowledge and skills required to navigate it, we have found that captains of advanced age and experience are the best leaders and managers of successful trading ventures. However, your extraordinary performance representing us and our captains as their mate and trading officer for the past four years has caused us to reconsider our prevailing practice.

"We have acquired another brig, the *Charles Doggett*. Accordingly, we have discussed your qualifications with four of your former Salem captains, e.g., Endicott, Kinsman, Putman, and Vanderford. Not only were they unanimous in agreement that you were extremely well qualified in navigation, but all also said you possessed the best all-around strengths of any with whom they ever have sailed. To a man, they agree you are gifted with cognitive abilities at the highest levels and a work ethic second to none. It is beyond ordinary reasoning and plausibility that you are yet to reach your 21st birthday.

"That said, we are prepared to offer you the command of this vessel along with a financial incentive and contract that will enable you to become a very wealthy man within the next ten years. We desire to acquire your services exclusively for our company not only to captain the *Doggett* but other ships in our fleet. We have the resources to make this relationship so mutually beneficial that you will never be hired away from us, nor would you want to leave. Accordingly, with no investment on your own, you will become a full partner sharing equally in the profits of every transaction. Additionally, you will have input on decisions as to our destinations, trading strategies, and purchasing choices. In other words, you will have free reign and full authority the same as an owner but without financial risk. Does this interest you, and do you have questions?" "Yes, I have two; where do I sign and when do we sail?"

All agreed we should continue to focus on the hot beche-de-mer market; so we planned to have our crew selected along with some repairs completed on the ship and to cast off for another voyage to the Fiji Islands by the first of April, 1824, and be gone for three or four years. As well, the timing provides the owners with some insurance advantages because I will become 21 on March 17th.

Our schedule was followed as planned, and on my birthday, our day for departure, a special event was held where the *Doggett* was docked. My parents and some friends, along with the Rogers Brothers, surprised me on the most unforgettable day of my life. I ceremoniously was awarded my official captain status and accompanying uniform, the *Doggett* was christened by its new owners and my mother and her friends honored me with a special American flag to fly on it. Eight states have been added since my birth bringing the total to the 24 displayed on my new flag. It was made of worsted bunting and measured 24 feet by 12 feet. Each state was mentioned aloud as we hoisted it for the first time. For such a glorious event, I named it *Old Glory*. It was the proudest day of my life.

It was made of worsted bunting and measured 24 feet by 12 feet

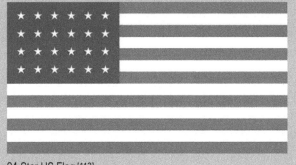

I named it *Old Glory*; it was the proudest day of my life.

Captain William Driver

24 Star US Flag [113]

Anchors Aweigh

It is difficult for me to describe the exhilarating feeling now that I am underway as the commander of my own ship. The potential to circle the world, visit cultures, people and places unknown, and particularly out of the reach of most folks, is at my beck and call as a twenty-one-year-old. I am awestruck and excited that the Rogers Brothers are committed to being second to none in the merchant marine business. They have challenged me to open frontiers that would advance their ambitions and contribute to mariners and their craft in general.

Since Salem is recognized as a worldwide mecca for sea captains and fleet owners, it is natural to expect that it would be a fertile environment and testing ground for the growth and development of ideas, innovations, and techniques that would advance the art of sailing. Necessity is the mother of invention, and a necessity led England's John Harrison to invent a marine chronometer that revolutionized navigation for mariners around the world.

There are numerous other necessities, and the Rogers Brothers have joined with other fleet owners to focus on one which is a major problem for all the seafaring—communications at sea.

Accordingly, I have added a new role and member to execute it on this voyage—a Communications Crewman. Other Rogers' ships and competitor ships will be working cooperatively to advance this capability. The rationale is that once separated from the land and alone on the vast oceans, communications outside the domain of our ship are next to impossible but for two options—signal flags and lamps. Both need much refinement and broader implementation, which will be a major focus on this three-year voyage.

Commonly in use since 1738 is Popham's Code, which is a reference book listing words, sentences, and numbers communicated by colored flags hoisted on the mizzenmast. For example, a knowledgeable signalman could read this message communicated by flags: [1]

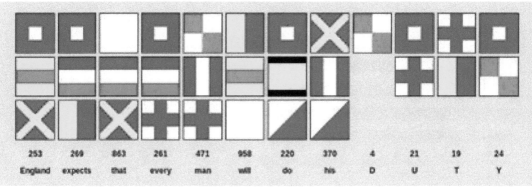

Popham's Code [114]

When two ships are passing in close proximity or can be viewed more closely through handheld telescopes, known as spyglasses, complicated messages can be sent back and forth. As well, handheld flags can be positioned and maneuvered by the ships' signalmen to change the messages in real-time. Whale oil lamps with shutters can be used at nighttime to send long and short blinks translated into coded messages. We also can "lie-to"—stop side-by-side—for voice messaging through megaphones. Flares shot into the sky provide another means of signaling messages indicating such concerns as danger and distress. Our immediate goal is to establish an improved messaging link to and from our Salem headquarters via the passing of ships to and fro. I believe a much simpler process and system is waiting to be discovered and I will be devoting time to this pursuit.

For the moment, I have other compelling thoughts. Talei is heavy on my mind. Our marriage was in accord with her tribe's tradition but not at all within the practices of my culture. I was given Talei as a present and was married without my intent or consent. Her father, Chief Asham, gave us a hut, so in the context of this unusual set of pre-arranged circumstances, nature drew us into an intimate relationship. Our affection evolved to love, and our separations even made our hearts grow fonder. By the time of my return voyage on the *Quill*, she had become fleshier which raised my suspicions.

Regardless, I look forward to seeing her again, but with some apprehension. Polygamy in her culture is engrained and reinforced by centuries of common practice. It is not viewed as immoral or uncivilized in the least. Conversely,

---

[1.] Similarly, Marryat's Merchant Vessel Signals also were used prior to 1857.

monogamy is the irregularity in her society and is looked upon as a sign of isolation, instability, and rejection. I learned that cultural inconsistencies present significant challenges especially when there are attempts to blend them. It is perplexing.

Having made this 24,000 nautical mile roundtrip previously, I have a good basis for what to expect—the unexpected. We hope to average seven knots and arrive in about 71 days. Since I have functioned in every role, operated every device, and conducted every task required of mariners, I am quite confident in overseeing the voyage to its successful completion…

Arrival in Fiji

After 72 days, at sea, we reached Talei's island where we anchored to join in on a welcoming celebration ashore. Running to meet our arriving utility boat was my smiling Talei, but not alone. By her side was a handsome bronze-skinned young boy along with an equally fine-looking adult male. She introduced the boy as Isaac, my son, and without uncertainties or reservation, she presented the man named Taito, as her husband and step-father to Isaac.

My initial ecstasy gave way to shock when we made our way to greet my old friend, Chief Asham. I had for him a customary present which will help overcome any obstacles to our ship's mission collecting more beche-de-mer. In this case, I brought him a shiny new flintlock pistol with a supply of balls, powder, and wadding. He grinned from ear to ear after taking a practice shot at a tree.

There was quite a feast and celebration underway on behalf of my return to the island's. However, I was not overjoyed with what appeared to be a first-class predicament with me in the middle. I seemed to be in a maze with no way out until I learned of a mitigating circumstance that placed this load-bearing matter in a different perspective. The tribe's guru was asked to explain the puzzle.

He responded, "the Island long-standing tradition, prevailing practice, and tribal law require that a child fathered by a non-Fijian be put to death at birth by strangulation. The preservation of racial and ethnic purity within our society is of the highest priority. However, it appeared that situation ethics came into play in this case. Isaac is the Chief's grandson and his favored daughter's only child. The resident witch-doctor and prophet using leaves and potions predicted that Isaac and his offspring were destined to produce descendants in very vast numbers who will migrate all over the world and bring great honor to the Fijian people.[1]

---

[1] Isaac's progeny are between their fifth and sixth generation in 2019. The earlier generations were very prolific with records commonly revealing 15 or more up to 24 children per generation. A statistical estimate using a conservative 105 children per generation projects there now are at least one million descendants of Isaac Driver. The prophet's prediction has proven to be accurate.

"Further, there was an aspect by which an exception could be made. I reminded the Chief that if a child is deserted by his father, and a surrogate father is willing to adopt him as his own and raise him in the Fijian ways, the child's circumstance of birth would not be the compelling factor, and the child's life would be spared. I argued that you, William, technically had abandoned Isaac even though unaware that he had been born. Further, I reminded the Chief that when a person's life is placed on the balance of justice, our laws require indisputable evidence if any factor is in doubt. I questioned if there is proof that you were the father and not Taito or someone else within our polygamous society. Further, Talei is willing to affirm that she, being at the age of her peak hormonal drive, experienced frequent intimacy during the many months you were together and since the two of you have been apart. At that point, the Chief exclaimed, "the boy will live!" William, this ends the story of the unusual set of circumstances that have enabled Isaac to survive and sit before you at this moment."

I thanked the guru for his thorough explanation and the Chief for his mercy. One incidental factor was conveniently ignored by all the participants who obviously were searching for a way to produce a favorable outcome; Isaac shares an undeniable likeness with me.

Like a lot of life's problems, time solves more of them than does wisdom. It is clear that Isaac must grow up in Fiji and in the Fijian ways. He will not be permitted to leave the island, nor is it a place of welcome for me to live. I plan to visit him whenever I am in the region and will support him with guidance and finances to help fulfill his dreams and aspirations.

Talei's situation is a different matter. Our cultures are so different and ingrained that they cannot blend. She is perfectly content, having two or more husbands, and I am not. Bigamy generally is frowned upon in the United States, and attempts are underway to make it illegal. Further, under the guidelines, provisions, and regulations of the matrimony laws, we are not married and never have been other than in Fiji. I will continue to be friends with Talei as I come to see Isaac, but not in a role or intimate manner as one of her husbands. So, this romance has ended—out of sight and out of mind. We will move on with our lives in our own individual and separate ways.

My trading procedures in Fiji, Manila, and later, Australia, have become a familiar routine, now in my third experience. We have established contacts and suppliers of completely processed beche-de-mer throughout the islands and several buying markets in Manila. The money from this voyage was compounding so fast we sent several large sums to Salem through the international bank in Australia. We were doing a complete turnaround from our curing sheds to the market every few months. Also, we continued to trade in the sandalwood for its oil whenever room on our ship would permit.

I ran into my pirate friends who were thriving as well. They even had purchased their own ship and were specializing in the tortoise shell market. They only purchase shells that remain as a byproduct of those harvested as a food source. This is extremely important in my value system, which respects the purpose of all creatures. I did some trading with them to take home a supply for the merchants specializing in jewelry, guitar picks, and ornament businesses.

On one of our buying excursions, we were exploring different territories looking for new sources of beche-de-mer. We were maneuvering our utility boat throughout the island sloughs using a thrusting pole against the shallow bottoms. Suddenly, we came upon a small settlement of English speaking men and women. We were welcomed ashore and learned that all were survivors from three different shipwrecks that sailed respectively from America, England, and Manila. They found food, water, and shelter in this setting of total isolation.

It was one of the most tranquil and self-sufficient environments I have ever seen. They offered us food and water and made one request in return. They implored us not to report their presence to the outside world. They explained that fate had delivered them to an uncharacteristic paradise free of worldly cares and woes. They had established their own form of government with elected leaders to administer and enforce law and order with minimal rules and intrusion into their individual freedom. They paid no taxes, the land and shelters were in

Leopard Tortoise [115]

abundance and free; they had plenty to eat and drink; there were huts and caves aplenty, and no one to bother them in this Utopia. It was a dreamworld—an example of how people ought to live in peace and harmony. The bubble soon burst when a quarrel erupted between two of them, and in seconds, the entire community exploded into a free-for-all. It seems that human nature does not change no matter where found.

Fijian Giant [116]

Hours later on a different island, we came upon another gathering with warrior-like appearances standing along the shore. I detected their language to be Fijian through which I was able to communicate a greeting, though seemingly not well-accepted. Standing amid this group was a towering man whose head rose at least a foot above the rest. He stepped to the water's edge muttering threats as the others retreated. I immediately recalled the introduction of Patagonian Giants given by Captain Vanderford on a previous voyage. This man minimally was seven feet tall, with quite a large head. Clearly, he was giant-sized and definitely foreboding.

All we need now to top the day is an encounter with some cannibals or pirates who are not at all strangers in these islands. Enough of this unproductive exploration; we moved on back to the safety and security of our anchored *Doggett*.

We had been in these islands for nearly two years and had just about depleted the supply of mature beche-de-mer. The harvesting areas need to lay fallow for an extended time to replenish, and we need to move on to Australia as planned. We bid farewell to all our contacts and especially Isaac. We embraced, and I told him I would try to visit often and bring him a special gift when I come. I left him a gold coin as a remembrance. Since there is no currency used on their island, and English ships are the most likely contact with the outside world, I left Talei a bag of pound sterling coins to pay for an unforeseen need that might arise on behalf of Isaac.

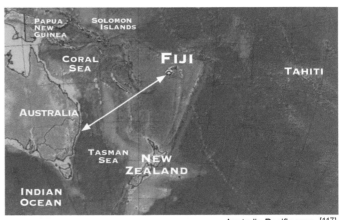

Australia Pacific map [117]

Australia

It is approximately 2,000 nautical miles and 10 sailing days from Fiji to Australia. Since it is our first voyage there, my owners want us to circumnavigate it to get a perspective on all the ports and trading op-

Great Barrier Reef [118]

tions. I learned from my studies and other captains that it is the only example in the world of a landmass being an island, a country, and continent, all in one by the same name. By being in the Southern Hemisphere, the seasons are reversed—hot during our winter and cold during our summer. The Tropic of Capricorn splits it across the middle at latitude 23°26'22" south of the equator; consequently, it is the southernmost latitude where the sun can be directly overhead. We found in our exploration incredible beauty and geological diversity throughout—deserts, lakes, mountains, ranges, valleys, and reefs.

"Four out of five animals and half of the birds that live in Australia cannot be found on any other continent on earth. Close to 140 species are marsupials which carry their young in pouches. Three mammals found only in

Australia actually lay eggs to birth their offspring. The country features some of the most poisonous snakes and spiders and the most dangerous bird (Cassowary) on earth. There are crocodiles up to 20 feet long and weighing up to 2,000 pounds that can live in fresh and saltwater. Along the coastline, great white sharks thrive in abundance up to 23 feet in length, yet more deaths occur from bee stings than all the above. For nature lovers, Australia is unmatched.

Cave Art Depicting a Giant [119]

As well, Australia has the world's largest coral reef system—the Great Barrier Reef—composed of nearly 3,000 individual reefs and over 900 islands."[93]

It seems the Patagonian Giant tales are following us on this voyage. The locals commonly share handed-down stories about the sighting of giants and unearthing skeletons approaching 9 feet tall.

The Aboriginals who originally populated the continent are believed to be a part of the first human migration out of Africa; Dutch explorers followed in the 1600s. The current British colonization started in the Botany Bay area in 1788. We will be arriving there at Port Jackson, finding the nation in the infancy of its growth and development, somewhat paralleling that of the United States. Other than farming, the primary businesses are the incarceration of criminals and the exportation of whale oil and wool. The first census reflected only a small minority of white individuals and a vast indigenous black population.

Aboriginal man and two companions [120]

It is peculiar how Australia initially grew its population. It was not by normal immigration but by indirectly serving as a penal colony for exiled criminals[94] from England. Those who were guilty of grievous crimes such as rape and murder never made it out of Britain. They were hanged or lost their heads in a guillotine; but those guilty of petty crimes such as stealing, disorderly conduct, defaulting on debts, and prostitution were exiled to Australia by the thousands. After seven to fourteen years, they were free to return home or stay and become citizens. Most stayed and embraced Australia as their new home. The descendants with whom we spoke when we went ashore joked about all their forefathers being criminals.[94]

Discovery Prison Ship at Depford [121]

We circled the island as planned and went ashore periodically to determine what products could be exported. The Rogers Brothers want me to evaluate if the country offers the potential to be included as a regular stop when we are trading in the South Pacific. I was just before ruling it out of consideration when a local sheep rancher opened my eyes. The very thought of wool awakes horrible memories of those course itchy trousers I was forced to wear to church services as a child. I was enlightened at the sight of thousands of sheep grazing across vast planes, and the incredibly intelligent shepherd dogs guarding and herding them by voice commands of their masters.

I learned the interesting history and jargon from the rancher. The word sheep means one or hundreds of sheep and if less than one year old, it is a lamb—a delicacy. The meat of an adult sheep is mutton, the same as for goats. It is useful information but not applicable to our specific trading interests. With nothing but frozen chunks of rapidly melting ice to preserve the meat, transporting it halfway around the world obviously is not an option. However, the wool they constantly grow is another matter—a sheep of a different color—as suggested by Shakespeare.

Merino Sheep [122]

Merino sheep render the most prized grade of wool in the world. Selective breeding in Spain and later refined in Australia have made Merino wool unique. It is comprised of the finest fibers and softest next-to-skin feel available. It is breathable, warm, odor-resistant and durable—no more itch and scratch. Indeed, the appeal for Merino wool makes Australia a very attractive option for future voyages.[95] Whale oil for lamps is a good secondary choice as well. I purchased a supply of oil and wool for analysis back in Salem. Perhaps of greatest significance is my purchase of a ram and two ewes along with hay and grain for the voyage. I know just the right rancher in Vermont who will love the opportunity to breed a domestic herd from the three.

CHAPTER 9
My Official Marriage

A Wet Sunday Morning [123]

The rounding of the Cape of Good Hope was like the home stretch in a horse race; it was smooth sailing back to Salem—a different world. Fiji and Salem are thousands of miles apart, but in customs, the differences are incalculable. The heavy influence of Puritanism in my upbringing, the Victorian standards of this time and place, my conversion to Christianity, and the lingering concern of having a child living in Fiji carefully must be examined, processed, and rationalized before I am comfortable asking Captain Babbage for his daughter's hand in marriage.

In the Spring, a fuller crimson comes upon the robin's breast;
In the Spring, the wanton lapwing gets himself another crest;
In the Spring, a livelier iris changes on the burnished dove;
In the Spring, a young man's fancy lightly turns to thoughts of love.
... Howsoever these things be, a long farewell to Locksley Hall!
Now for me, the woods may wither, now for me the roof-tree fall.
Comes a vapour from the margin, blackening over heath and holt,
Cramming all the blast before it, in its breast a thunderbolt.
Let it fall on Locksley Hall, with rain or hail, or fire or snow;
For the mighty wind arises, roaring seaward, and I go.[96]

—Locksley Hall, Alfred Lord Tennyson

I have courted the lovely Miss Martha Babbage for several months since returning from Australia. I had met her through her father and strictly have followed the conventional Bay Colony customs in terms of respect, decorum, and dating etiquette. For the first few months, our contact in her home was always in the presence of an adult relative; and when taking a stroll or attending an event, a chaperone was closely behind. Holding hands and a goodnight kiss took months to be tolerated by our supervisors, and even then it always was in the presence of a witness—a huge departure from the practices found in the islands.

We never were left alone, but our chaperones graciously would give us some space to converse without being overheard. In time, we became serious to the point of discussing marriage, so I decided to end the indecision about the nagging concerns. I applied Occam's Razor principle and utilized the approach that would be the simplest and have the fewest variables. Honesty and full disclosure were my choices regardless of the outcome.

I thoughtfully exposed the negatives of her being a stay-at-home wife of a seaman. Not only will there be months of separation, sometimes it will extend into years. I told her about Talei, Isaac, and my entanglement caused by the customs of the Fijian society. Lastly, I admitted that I had faltered and failed in the conflict between nature, my immaturity, my value system, and strength of character; I had asked for God's forgiveness and believed I received it. Relieved that I had been forthcoming, I asked Martha—in spite of the foregoing—if she could accept me as her husband, she would have my solemn promise to provide her needs, be a good husband, and be faithful until death do us part.

Martha grasped my hand while smiling and said that she understood the separation factor quite well; the years living in the home of her seafaring father was an adjustment, but a rewarding tradeoff for financial independence and early retirement. Furthermore, she wanted children—lots of children—to nurture and occupy her time. Because of my honesty and the unusual circumstances that implicated me in Fiji—she forgave my transgression, accepted my remorse and pledge of faithfulness. I was relieved and reassured that my approach had been the right one; honesty always is the best policy. Within weeks, I received her family's approval, and we were married, found a home, and began our life together.

The Rogers Brothers were considerate of me being in a new marriage and switched me from the *Doggett* to the *Clay* and *Black Warrior* for several shorter West Indies round trips. This happened off-and-on for four years, which enabled us to solidify our marriage and produce three beautiful children.

A Stormy Reunion

On January 13, 1831, I was reassigned to the *Doggett* for a year-long voyage to New Zealand, Tahiti, and an unplanned destination, which was a shocking surprise to me. On the fourteenth, my crew and I departed in moderately threatening conditions for the first leg of the journey headed south toward the Horn and the Pacific Ocean. Oddly, the wind which had been blowing briskly from the northwest for several days temporarily lulled and became light and puffy in a clear sky—the quite familiar "calm before the storm." There was nothing unusually threatening except for a darkening sky far to the north. We had in sight six other vessels of various rigs, all outward bound. They were steering for the South Channel exit around Marblehead which was the nearest and natural. However, they confronted the tempest at its fullest fury, and all vessels and their crews perished except for the *Statesman*. She attempted to turn back but was torn to pieces on Middle Island near point Alderton. Fortunately, her crew was near the shoreline and most survived.

I was at the wheel and chose to exit out the East Channel that would

Shipwrecked [124]

enable us to clear the most perilous area and then turn south. I encouraged our frightened crew to remain calm. I did not share my growing concern that a snowstorm off in the distance appeared to be imminent and would be our next challenge. I thankfully credit my experience and training for helping us overcome the first hurdle. At 7 p.m., Thatcher's Island Lights were in sight some twelve miles away. We were in open water but with a cold north wind building rapidly. I struggled with the weather wheel as the spray from the pounding hull was freezing in mid-air and on all our faces making us dreadfully miserable. The eye of the storm was overtaking us, unlike anything I had ever seen. Our poor men, barely away from their homes and firesides, were scantily clothed and unprepared physically and emotionally for such a crisis. We battened down our fore

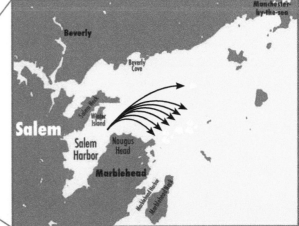

Seven ships leave Salem Harbor, six sink in the storm[125]

hatchway and took those suffering the most below into the cabin for a degree of warmth and dryness provided by a small stove.

The bitter cold, heavy fog, turbulent sea, and biting winds followed us through the night. At 4:30 a.m. the *Doggett* labored heavily, and at 5:30 a.m. we lowered most of our sails and were running virtually free.

The Tempest [126]

Our brig was rising, shaking, plunging, pounding, and groaning relentlessly when a crewman loudly exclaimed, "she can't stand it, sir." I responded that with the deadly shoals beneath us, it is better to flounder here without one board remaining than to bury our bones in the fatal sand.

With a couple of sails partially engaged and moving us at about 4½ knots, we turned east to northeast into the face of the arriving snowstorm. The conditions progressively worsened until the decks became knee-deep in slush. I ordered all remaining crew to the cabin below, except for the second mate, Winn, along with crewmen, Francis and Smith, who were manning the wheels. Charley, my huge mastiff, was terrified in the cabin below, nudging closely to anyone who would offer a comforting pat or cheerful word. Red, the ship's good-luck rooster, clung tightly to a boat rib as he perched out of the direct wind beneath our utility boat. Continually, every time the blizzard seemed to be getting the upper hand and our spirits were beginning to sink, old Red would cock his head and defiantly crow as though he was challenging the storm to a duel. His cockiness seemed to give us an adrenalin rush that helped us continue battling.

Oh, it was truly a horrible night for man or beast—one to try our souls and fortitude. I stood on the creaking deck, helpless but for God. I prayed for wisdom, courage, and the strength to make good choices that would comfort and safely deliver my crew. It was a scene which no pen can describe as the cold breath of death tried to chill our determination and particularly my resolve. A despairing gaze in the eyes of those worried seamen constantly scanned my face for a glimmer of hope, a cheerful look, or a word of encouragement. They seemed to ask, "can we make it, Captain, do we have a chance; what shall we do?"

They sailed. They sailed. Then spake the mate:
 "This mad sea shows his teeth to-night.
He curls his lip, he lies in wait,
 With lifted teeth, as if to bite!
Brave Admiral, say but one good word:
 What shall we do when hope is gone?"
The words leapt like a leaping sword:
 "Sail on! sail on! sail on! and on!"

Then, pale and worn, he kept his deck,
 And peered through darkness. Ah, that night
Of all dark nights! And then a speck—
 A light! A light! A light! A light!
It grew, a starlit flag unfurled!
 It grew to be Time's burst of dawn.
He gained a world; he gave that world
Its grandest lesson: "Sail on! sail on! sail on! and on!"[97]

Columbus—Joaquin Miller

[127]

The morning broke, revealing a mix of muddy, green, and yellow water all around us in the midst of a wild and confused sea. We sounded the depth and found it to be seven fathoms over a gravel bottom, and again it was five fathoms continually growing shallower as we drifted toward the shore. It was such that we knew in less water we would be lost. However, the wind veered a little northward; we put the whole foresail to work, and it kept us from grounding. The storm seemed to settle and was so heavy and damp it appeared to break up the sea a bit and nudged us into the Gulf Stream. We hoisted all our sails and *Old Glory* as we pulled away for New Zealand.

When I consider that the *Charles Doggett* is the only ship to survive out of the seven that departed Salem yesterday, I cannot help but wonder why. Indeed our fervent prayers were answered, but somehow, I believe God has a special mission and purpose for William Driver; and if called, I pledge to respond with all my heart, soul, and energy for as long as I live.

All our animals, including a load of cattle, perished in the storm but my dog, Charley, and Red, my rooster. They happily joined us on deck for a devotional service of praise and thanksgiving, a practice that will happen every Sunday from this day forward on any ship which I am in command. With shaky voices, hands lifted, and tears flowing down salty cheeks, we closed our service with this hymn:

Praise God from whom all blessings flow:
Praise him, all creatures here below:
Praise him above, ye heavenly host:
Praise Father, Son, and Holy Ghost.[98] Amen

Bay of Islands, New Zealand

After surviving the first storm, another moved in and joined with the turbulence at Cape Brett where the Tasman Sea meets the Pacific, north of New Zealand. I checked on matters below deck and found that our stove had turned over and dumped burning coals near to the kegs of explosive powder. The coals and stove were quickly removed and joined Davy Jones at the bottom of the sea.

I returned to the weather wheel to find Smith needing relief from frostbitten feet. His socks were soaked and boots filled with icy slush; yet he had remained at his post unyielding. I took over the wheel in my sock feet and gave him a pair of dry socks and my boots. I remained relieving our two helmsmen at the lee wheel and weather wheel for the entire five hours it took us to round Cape Brett. The first storm challenged our resolve, and the second only strengthened it. The seas and skies finally relaxed and were so incredibly peaceful that we wondered if we had just awaked from a bad dream—a nightmare.

We arrived at the Bay of Islands on June 5, 1831, some one-hundred and forty-one days since departing Salem. Our anchor was dropped in the harbor of Kororaeka[99]—the first permanent settlement and seaport in New Zealand. There were a dozen or more ships in the harbor because of the strong whaling interests as well as for the onshore attraction provided by hundreds of loose females. In fact, Kororaeka became the largest whaling port in the Southern Hemisphere, and with its flourishing prostitution trade along with its widespread lawlessness, it earned the reputation as the hell hole of the

New Zealand relief map [128]

Pacific. To say the least, we had some reservations and grounds to be cautious. Our focus was as a business venture purchasing whale oil for lamps and wax for candles. As well, we needed to repair minor damage from the storm and replenish our water supply.

However, before having time to think about trading interests, a very bold native thieving party led by the terrible Tia Tora, crowded on deck the first morning of our anchorage. They numbered at least 400 and started taking almost everything they fancied.

I told Tia Tora, in his native language, to get his men off the ship immediately as our crew drew their pistols in a show of strength. He only laughed and mocked us as his men continued to steal. I noticed they kept their distance from Charley, my huge mastiff weighing over 200 pounds and tied to a mast pole. He was as big as a calf and was pulling at his constraint with hair bristling on his back and growling at the thieves when Pablo, our cabin boy, cut him free. With the fury of a lion, Charley lit into them tearing flesh right and left as they started running into each other, dropping their prizes and diving into the water while howling like wolves. Their leader start-

A Charley and his Pony Friend [129]

ed hopping and dragging with Charley clamped down on his leg. Blood was streaming everywhere as he jumped overboard to join his fleeing band. Thereafter, I was known as the Captain with the big dog.

The next morning we sailed to a missionary station across the bay at Paihia for water. I met and told the founding missionary, Rev. Henry Williams about our encounter with the thieves. He was greatly surprised that I was still alive. A year before, an English ship had been boarded by the same party, and all were killed and eaten. I was anxious to learn about the mission's work and to share about my recent spiritual renewal experience.

However, the Reverend and I did not continue our conversation cordially once he learned I wanted to replenish our water barrels from his mission stream. It is a courtesy countries worldwide typically extend to visiting ships as long as the supplies are endless or abundant. He wanted an unreasonable trade for goods in exchange for the water which I could not afford. I offered a local native chief who had access to the stream a musket to intercede for me and fill my barrels. This action led the Reverend to have his boys cut trees across the entrance and add a fence to block others in the future from sharing "his" water—from a running stream.

Rev. Williams characterized me as being obstinate and selfish; I found him likewise, along with being the most out-of-character representative of the ministry I ever met. It was such a trivial matter, but I am confident the incident will cause a future loss of commerce for the community and contributions to the Mission. Once back in Salem, our fleet and many others will not take kindly to such abnormal hospitality. There are many other options where merchant ships are most welcome, and their water is free.

Once underway and having cooled down a bit, I asked God to forgive him and me.

> "And whosoever shall not receive you, nor hear your words, when ye depart out of that house or city, shake off the dust of your feet."[100]
>
> Matthew 10:14

Tahiti

We remained at the Bay of Islands until the 26th of June and departed for several stops en route to Tahiti. We had acquired a good supply of whale oil and wax in New Zealand and some beche-de-mer in some of the Fijian islands. A highlight for me was a stopover at Fiji's main island for a visit with Isaac. We spent half a day together, updating and enjoying each other's company. I gave him a pocket knife for his surprise, and as we departed, promised to continue making visits every time I am in the area. I did not see Talei.

Tahitian Girls [130]

We arrived in Tahiti 17th day of July—an incredibly beautiful island with strikingly attractive people.

We were met in the harbor of Matavai Bay by an exceptionally impressive and well-formed young lady escorted by two powerfully built men in an island dugout. She politely requested permission to come aboard, which I obliged. I had heard how pretty and cordial their woman were, but was surprised when she introduced herself, in near-perfect English with a British accent, as Queen Pomare IV.[101] She had been the queen of Tahiti since 1827 and succeeded as ruler at age 14 after the death of her brother Pomare III.[102] She was very impressive in that she seemed to have vast knowledge about the world around her even though she had never left Tahiti. I assumed that she had learned much from her contact with ship captains and crew members that visited Tahiti from around the world, especially from England and America. As well, I introduced myself—not staring—but unable to avoid noticing that she was bare to the waist except for long strands of shiny black hair hanging to her mid-section. She brought me a welcome bouquet to celebrate my first visit and after that offered her permission for us to anchor; I had a trinket for her. We exchanged pleasantries and shared backgrounds finding a common passion for the oppressed. Her given name, Aimata, means in her language "eye-eater" in keeping with an old custom whereby rulers ate the eyes of defeated foes. In her experience as the Queen, her knowledge and refinement about the outside world were very impressive. The famed Charles Darwin was one of the many dignitaries she has hosted during her successful reign. She asked for time to discuss a troubling matter with me in private. I agreed and ordered tea to be brought to my quarters,

Queen Pomare IV [131]

which suited her as long as her guards could remain stationed just outside our closed door.

She began the conversation by asking if I was familiar with Captain Bligh and the mutiny of his ship back on April 28, 1789. I grinned and told her that I had heard about several mutinies but not that specific one; it occurred before I was born. "Neither was I alive at that time," she replied, "but I have been troubled by the result of it since I became Queen. It is important that you understand the story."

The Mutiny in Retrospect

The Queen continued with her account. "Captain Bligh was an officer in England's Royal Navy and was commissioned by King George III to transport breadfruit plants and trees to the West Indies. Ultimately, they were to provide a low-cost source of food for the growing slave population. The plants happened to be in abundance here in Tahiti which is the reason Captain Bligh and his ship, the *Bounty*, spent five months here collecting and preparing to transport over one-thousand of the plants.

"Now, Captain Driver, we are grown-ups and can be straightforward when discussing human nature. I have learned much about your conventional culture as a result of the frequent merchant ship visits from America and Europe. It is unlikely that you are as fully aware of mine. When you come ashore, you will observe a cross-section of people of mixed appearances and skin colors ranging from black to white and all shades in between. Simply put, sexual intimacy here is practiced freely and viewed as normal without restrictions or formal bonding ceremonies. Consequently, many of the men women and children you will see are the offspring of the original *Bounty* crew. Since your men have been at sea ten months and plan to be here for an additional five, you should not be shocked to find similar behaviors unless you keep them in chains and locked up.

"Now, back to Captain Bligh and his men. As I learned directly from eye witnesses who still are living on this island, the five months spent here by the original Bounty shipmates were likened to Heaven on earth and Utopia combined. Therefore, when it was time to finish their mission delivering the breadfruit plants to the Indies, there was a great reluctance for the crew to leave. Three tried to desert but were flogged with 21 lashes from a "Cat O' Nine Tails."

"Begrudgingly, the rest followed Captain Bligh's orders and sailed away; but just two weeks out of Tahiti, rumblings of discontent grew louder and louder until on the morning of April 28, the somewhat cantankerous commander was waked up and bound by a mutinous crew led by Fletcher Christian, the second in command. In spite of giving dire warnings to the mutineers, Bligh and eighteen of his loyalists were set adrift in the ship's 23-foot utility boat with only enough food and water for five days. For some unknown reason, they were permitted to take with them Captain Bligh's personal compass, navigation tables, a quadrant, a broken sextant, and his pocket watch. Arguably, those items would be of no help in maneuvering the craft, so logically it was assumed by friend and foe alike, they were doomed to die and take all evidence of wrong-doing with them to the bottom of the ocean.

"Captain Bligh was known to have a volatile temper, oppressive manner,

Captain Bligh and his Loyalists [132]

and inclined to administer harsh punishment. However, most believe there was only one compelling reason that instigated the mutiny, and it had nothing to do with Bligh's personality and traits. It was the relationships with the tender and charming Tahitian maidens, along with nature's powerful attraction that were the catalysts. Many fell as deeply in love as any courtship could create in any society in the world—with or without a formal ceremony.

"Strangely, as good and bad luck always have their way, 47 days later the rejected Captain and his loyalists found their way to Timor in the East Indies—a British colony some 4,000 miles from Tahiti. A year later they were back in England.

"Concurrently, Fletcher Christian and 25 others returned on the *Bounty* to Tahiti where 16 wanted to remain. They thoroughly understood that the authorities likely would be looking for them here. Christian and the remaining eight, along with six Tahitian men, twelve Tahitian women, and a child decided to leave in search of a hide-a-way somewhere in the South Pacific. After many days, they found a very isolated uninhabited volcanic island 1400 miles east by the name of Pitcairn. There I learned, they sunk the *Bounty* and colonized the island.

"As you might expect, the Royal Navy was none too happy over the theft of one of their ships and, in particular, the at-large mutineers who took it. So, the Admiralty sent Captain Edwards on the *HMS Pandora*[103] to the South Pacific in search of them. The mutineers who returned to Tahiti were captured by Cap-

tain Edwards and returned to England for trial; three were hanged. He never found Fletcher Christian or Pitcairn Island.

"In the after years, there were several sightings of Christian's group living on Pitcairn. The *Topaz,* an American whaling ship, commanded by Captain Mathew Folger came across Pitcairn Island on February 6, 1808, and discovered the settlement—19

Pitcairn Island [133]

years after the mutiny. Captain Folger sent news of his findings back to England, but being involved in the Napoleonic Wars, their existence was ignored.

"Some six years later, two British ships, the *Briton* and *Tagus*, discovered the island and its community on September 17, 1814. Captains Staines and Pipon found one mutineer, John Adams, was still living. Fletcher Christian had been killed, and the remaining community was beset with illnesses and famine. Although Adams was considered to be a fugitive, he favorably impressed the captains by the example he set and his testimony; so they agreed it would be 'an act of cruelty and inhumanity' to arrest him. No doubt, they convinced the British Royalty of the appropriateness to move the colony. On February 28, 1831, the *HMS Comet* and the Colonial Barque *Lucy Anne* arrived from Sydney and removed the community of sixty-six men, women, and children back to Tahiti. They arrived back here on March 6, 1831.

"My people treated them with great generosity and kindness, providing food and housing. However, they did not feel at home with the honor and dignity as they did before they left.

> "A prophet is not without honor except in his hometown and in his own household."[104]
>
> Matthew 13:53–58

"Considering their European cultural background and strong religious beliefs, they simply cannot adjust to our liberal ways, particularly as relates to dress, intimacy, and morals in general. They have lost their immunity to our diseases, and many died. Within a month of their arrival in Tahiti, Thursday October Christian I, the son of Fletcher Christian and who was the first child born on Pitcairn, died. His death was followed by the youngest to be born, Lucy Anne Quintal; and during the next two months, there were 10 more deaths and only a single birth.

"Now, to the point of our discussion, Captain; they and I need a big favor from you. Please take them back to Pitcairn."[105]

The Queen took me to see them huddled in a large grass hut—the most pitiful collection of humanity I had seen since the lepers in India. They were brokenhearted, hollow-eyed, and forlorn—as helpless as humans could be. How could they escape? I saw they would be lost if left where they were. I told the Queen that I did not have authority from my employers to make changes in my itinerary. To do so would cost them lots of money; yet, they had instructed me to always use my best judgment in unforeseen circumstances. The people cried with extended hands, and the Queen begged. They offered all they had of value—their blankets and shells worth only pennies. At most, the combined value would be about $500—a paltry sum compared to what that decision would cost my owners and me. Although the owners are wealthy, their loss would be in the thousands of dollars, and mine could cost me my brig and possible life. I pondered…

> …he saw a very poor widow dropping in two little copper coins. He said, "I tell you that this poor widow put in more than all the others… [they] offered gifts from… their riches; but she, poor as she is, gave all she had…[106]
>
> Luke 21:1-4

God spared us in that dreadful storm when six ships had perished with their crews. Was this the mission for which my life was spared? I sailed from Papeete, Tahiti with 65 passengers on the 14th day of August, 1831, and after 1400 miles of perfect weather, delivered God's children to Pitcairn Island on the 3rd of September, 1831. If I never earn another penny, their smiles and gratitude

Note of Gratitude

Pitcairn Island, Sept. 3, 1831

This is to certify that Captain William Driver of the brig *Chas' Doggett* of Salem carried 65 of the inhabitants of Pitcairn Island from Tahiti back to their native land, during which passage Captain Driver behaved with the greatest kindness and humanity becoming a man and a Christian, and as we can never remunerate him for the kindness we have received we sincerely hope (that through the blessing of the Almighty) he will reap that reward which infallibly attend the Christian.

Signed: George H. Nobbs, Teacher
Arthur (his mark) Quintal
John Buffett
John Evans

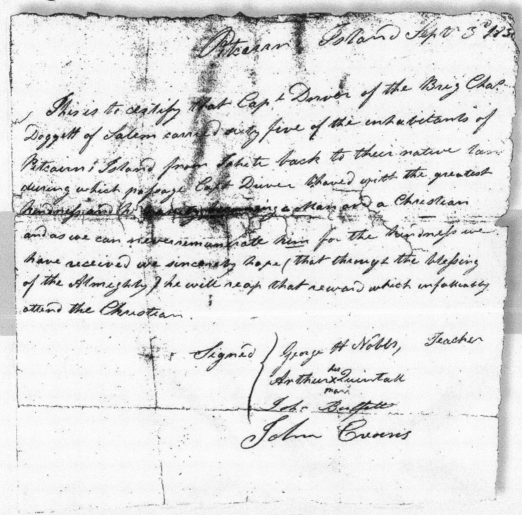

Photocopy of original Note of Gratitude given to Captain William Driver from Pitcairn people.[134]

have made me a millionaire.

As I departed with a lump in my throat and tears in my eyes, their leaders handed me this note of gratitude.

After leaving Pitcairn, we refocused on our trading mission between Fiji and Manila. We acquired 1600 pounds of premium turtle shells, bundles of bows and arrows, war clubs, and other primitive warfare artifacts. The shells alone were worth over $20,000—a considerable fortune in 1832.

En route back home, we made a couple of stops in the West Indies and found some unusual bargains in Cuban cigars and Jamaican sugar. Strangely, back in Salem, our profits were as much as they would have been without the detour.

The Rogers Brothers were delighted that we still had a good profit after hearing about the episode; they remarked that the tale would make a good book.

We arrived in the late spring of 1832 to find another assignment awaiting me on a different ship, the *Black Warrior*. It was good to be home for a few months spending quality time with Martha and our three children, William Christopher Driver, Eben Ropes Driver, and Martha Silsbee Driver. Naturally, Martha wanted to hear about all the places I had visited and the adventures involved. The encounter with Tia Tora made her skin crawl, and the Pitcairn mission brought a tear and smile.

In retrospect, it is with the greatest humility and thankfulness that I found a niche in the merchant marine trade. Although it is physically and mentally challenging and is very high risk—often to the degree of being life-threatening—the rewards come fairly quickly and in favorable sums.

That said, I am considering retiring from the earning phase of my life to the returning phase—to give back to society in some form or fashion what America's incredible free enterprise system has provided me. Perhaps it will be in a spiritual endeavor or a public service role. My savings will supplement other jobs I will find for me to support my family and serve my country as well. I am not ready immediately, but I am moving in that direction.

The *Charles Doggett* is dry-docked while undergoing repairs and restoration of the damages caused by the storms during the previous voyage. Meanwhile, I was placed in command of another Rogers Brothers ship, the *Black Warrior*. My trips were of shorter durations and mainly to the West Indies.

However, the Rogers Brothers are beginning to increase their focus in more distant markets such as Australia. In recent trades there involving cattle and kangaroo hides, the *Tybee* alone transported almost 6,000 pieces to Salem. Accordingly, I became involved and arrived with the *Black Warrior* at Sydney in October of 1833 to find the country in near-famine conditions. They desperately needed flour. I sent Joseph Rogers, my First Mate, back to the States to procure a full load while I remained to set up a depot for storage and exchange.

He returned in August 1834 with 1600 barrels which I ultimately marketed for almost $20,000 in profit.

While I was occupied with the flour sales, I sent Rogers to the Bay of Islands, New Zealand to trade for whale oil. This led to several other round trips to the States by Rogers while I remained trading the flour and accumulating more barrels of oil to store in our depot warehouse.

Now in late 1836, while returning to Salem transporting thousands of gallons of oil, I find myself in a reflective mood. This three-year venture alone has parlayed the earnings for the Rogers Brothers into immense financial gain, and me, financial independence. Strangely, my passion for the sea is waning. The grind of this lifestyle is taking its toll.

Nearing the harbor entrance at Salem, I had an eerie feeling—a premonition of finality. Could this be my last voyage? Has the salt lost its savor—still in my early 30's?

My answers soon came into focus and a roadmap clearly in view. I found my Martha to be extremely ill with throat cancer. She had been struggling for months to tend to matters at home and particularly our young children. We have only a few elderly relatives remaining in Salem, and they barely could take care of themselves. Martha's condition worsened daily, and on September 5, 1837, she mercifully passed on to her heavenly home. My world stopped turning.

> How wonderful is Death,
> Death, and his brother Sleep!
> One, pale as yonder waning moon
> With lips of lurid blue;
> The other, rosy as the morn,
> When throned on ocean's wave
> It blushes o'er the world;
> Yet both so passing wonderful![107]
>
> Queen Mab – Percy Shelley

CHAPTER 10
My Last Voyage

The Tide Rolls in [135]

The tides of life have brought me ashore to a crossroads moment. My ties to Salem, Massachusetts sadly have ended with the passing of my wife. My roots here are deep—over 200 years[109] since my forefathers came. It is not easy to disregard such a strong and enduring connection to the Bay area. If I stay, the same ocean breeze that waked me as a baby still is in charge and will continue to entice me with every breath of its briny air.

My mate's absence from the familiar setting and routine surely will continue to depress us as we struggle with our grief—especially so for my young children. An empty chair, a soft voice, and a missing word of comfort only a mother can give, create a void they soon cannot forget or understand. The melancholy from the constant reminders is too much for them to bear, and the burden is too heavy for all of us during this transition. The passing of time will help, but it seems we need to distance ourselves from the vividness of things which conjure memories that linger and haunt us.

I labored over my options concerning a different locale for many weeks. If my decision only affected me, it would be relatively simple and indifferent; but in this case, the lives of three young children and their progeny for generations to come will be impacted and long-lived. Therefore, the balance tips in favor of what I believe to be in their best interests. I strongly have considered moving to Nashville, Tennessee, to join in the family circle and support systems of my brothers and their families. I studied the nature of the territory and its interesting history closely.

Tennessee was admitted to the Union as the 16th state in 1796, but Nashville actually was founded seventeen years earlier in 1779 as Fort Nashboro. Originally it was Indian Territory on land ceded from the State of North Carolina. It is located on the bluffs of the Cumberland River and was founded by pioneers John Donelson and James Robertson. In 1806 it was chartered as a city. I am older than its charter by three years.

Of particular interest to me as a seafarer is the city's reputation as a major inland trading port. Amazingly, a boat launched on the Cumberland with nothing but a push pole will flow downstream from the city westward past Clarksville, Tennessee, then due north across Kentucky to the Ohio River. After that, it will follow the same water trail taken by Lewis and Clark while exploring the 1801 Louisiana Purchase. Fifty miles farther westward, the boat will enter the Mississippi River for the final leg due south to the Gulf of Mexico at New Orleans, Louisiana.

I studied and evaluated every aspect of relocating and have been counseled by all my Nashville kin—broadly and deeply. My conclusion is that God, family, education, religion, living conditions, traditional values, and economic opportunity combine to make Nashville a good fit for us. I told the Rogers Brothers of my plan to retire from the *Charles Doggett* and to take with me only two items—my trunk and my beloved flag, *Old Glory*; it has flown on all ships I have commanded since I was 21. They reluctantly accepted the news but wished me well, saying a job always would be waiting should I want to return.

Fort Nashboro [136]

In many respects, Nashville mirrors Salem in its infancy. Just like their grandparents, it will provide my children the opportunity to be pioneers in the early growth and development of a fairly young frontier town. My brothers, Henry and George, along with their families, praise living there because it features a moderate climate, a wholesome environment, and major economic opportunity. They followed the cobbler trade the same as my father and went to Nashville to establish a wholesale shoe business. Both brothers have offered us the option to live with them until I get settled, and the children are back in a comfortable routine.

We arrived in September 1837 and accepted Henry's offer to join his family since his home is within walking distance of several churches, a doctor's office, a grocery store, the wharf, and the main business district. A bonus included his wife, Ellen, and her niece, Sarah Jane Park, who were eager to provide home-schooling and nurturing for my children. This factor is especially significant because there are no public schools in Nashville. Fortunately, my savings from twenty years at sea is such that I do not have to work to pay my expenses. Henry will not accept rent, but I insisted on volunteering to help him in his business and paying a generous weekly stipend to Ellen and Sara Jane. Their support and services sustained us during this especially crucial time in our lives.

William, Eben, and Martha, my children, really were smitten by Sarah Jane.

She quickly earned their respect and affection almost as much as they had for their birth mother. They accepted her guidance, teaching, and discipline without a bit of resistance.

Sarah Jane never talked about her mother and father, and as a matter of local etiquette; I did not ask. She loved being around her Aunt Ellen and Uncle Henry. When not tending to the children or doing chores for Ellen, she would be at the shoe store running errands and deliveries for Henry.

Sarah Jane was yet to be sixteen though she looked like she was at least twenty and had the maturity of a thirty-year-old. Perhaps not often being around girls her age and associating only with ladies twice her age, she naturally acted like them. On occasion, we would take the children grocery shopping or on an afternoon picnic at the river park. It was quite common for strangers to confuse us as husband and wife.

In a short time, it became clear that we were developing a mutual attraction. I was quite mindful of our age difference and was most careful to respect that fact. However, we emotionally became closer each day until a time when she reached for my hand while we were walking, and on another day, my arm while we were talking. Due to my hesitancy and caution, she ultimately broke the ice by suggesting that I needed a wife and the children needed a mother and asked would I be interested in her filling those roles—a very wise approach. I did not respond directly but asked her about how she felt about our age difference and me having already been married. [I never discussed Talei and Isaac with anyone but Martha and do not intend to ever resurrect them from my past.]

Sarah Jane responded from her strong religious background and frame of reference in a most compelling way. She remarked that age gaps of ten, twenty or more years are quite common in her network of friends and acquaintances. The life expectancy is less than 40 years in Nashville at this time in history, so there are lots of widows and widowers needing mates to fulfill the purpose of marriage in the first place—conceiving and rearing children.

Additionally, she had heard sermons on the subject and had studied the scriptures herself to find the only mention of an age requirement is an implied minimum being at puberty and with no maximum. Abraham was ten years older than Sarah and Joseph significantly older than Mary. Boaz was much older than Ruth, and she was praised for not seeking a younger man.

Sarah Jane summarized her reasoning by saying we had met all the conditions and avoided all the prohibitions—spiritually, biologically, and logically. Therefore, our love for one another is the only remaining consideration for us to weigh. When we moved to Nashville, I never had a thought of marrying again, but on January 13, 1838, at Christ Church Cathedral on High Street, we were joined in holy matrimony by Reverand John Thomas Wheat. My children were present and were very much in favor of this union. From that day forward by their own choosing, they always addressed Sarah Jane as "Mother."

After arriving in Nashville, we bought a two story house on South Summer Street and paid cash in full from my savings. There are numerous trees on our lot and in the neighborhood. I immediately prepared to display *Old Glory* in a prominent place outside to inform the neighborhood an American Patriot has moved in. It will be strung across the street to celebrate special occasions and properly displayed. My son, Henry, got a big thrill each time he would climb the locust tree and attach a line. We then fixed a single wheel pulley and ran the line to our upstairs window where we could adjust the flag at any position over the street. It was an impressive sight and drew lots of admiration throughout the neighborhood. It became a real landmark in Nashville as it was flown on almost every holiday and major occasions including the Fourth of July, St. Patrick's Day and George Washington's birthday, which also happened to be mine.

Our new residence was plenty large enough to store all of the artifacts I col-

Rope for *Old Glory*

William Driver Home "My house" on South Summer Street in Nashville [137]

lected from ports around the world. I brought furniture and other household goods from my former home in Salem. Included in the items I brought from the *Charles Doggett* was my old sea chest that my children named "Father's Ship Locker." It was a leather-covered trunk sprinkled with brass tacks and contained my camphor wood chest in which *Old Glory* always was stored when not in use.

Our house is a lovely place by the side of a main road surrounded by a picket fence. There is a yard swing under a shade tree where we can relax and watch the people go by. It has an upstairs with several bedrooms, two fireplaces, a deep water well, a cistern, and a two-hole outhouse—all the luxuries a family could ever want.

I lived by the sea in Salem watching the ships come and go; now by the side of the road, I watch a parade of humanity go by. Some travel by horse and buggy, but most walk and frequently stop to chat. These Southerners are a friendly lot. All of them speak to friends and strangers alike while passing, and the gentlemen courteously tip their hats when greeting a lady. It is entertaining to watch them and listen to their comments. There are all kinds—mostly good people and maybe some bad; but are they not just reflections of us all?

"I see from my house by the side of the road, by the side of the highway of life, the men who press with the ardor of hope, the men who are faint with the strife...

I turn neither away from their smiles nor their tears—both parts of an infinite plan.

...Let me live in a house by the side of the road where the race of men go by; the men who are good and the men who are bad, as good and as bad as I.

I would not sit in the scorner's seat nor hurl the cynic's ban—

Let me live in a house by the side of the road and be a friend to man.[110]

—Sam Walter Foss—*The House by the Side of the Road*

Nashville waterfront [138]

In the coming months, life settled down for us, and a comfortable routine was established. Sarah Jane and the children were enjoying each other's company spent with schooling, hiking, and watching the rapidly growing steamboat commerce at the nearby wharf. The children were as excited as Sarah Jane and I when she announced that another brother or sister would be joining us within the year.

I volunteered at the shoe store, but it was not prospering due to the lack of customer traffic. Nevertheless, I was able to relieve Henry for his other pursuits, and it was important for me to be occupied with something productive. However, it ultimately became unproductive, and before long, the business failed. Henry moved on to other options, and I began to focus on personal interests in the public good.

CHAPTER 11

The Abolitionist

In my new lifestyle, I am following the advice of many sages who were examples for me along the way. They all agree that happiness follows those who remain active in endeavors where other people depend on us every day—doing things they are incapable of doing for themselves. It might be building the morning fire in a poorhouse, enabling the disabled, tending to the sick in a hospital, ministering to the flock in a church, or being an advocate for those treated unjustly by their government—anywhere others need us to be their surrogate. Conversely, they warn, if we fail to contribute to society, our ability to serve will be taken from us as well as our usefulness; it will follow that we will wither and rust from inactivity. St. Luke, in his parable of the pounds, gave a parallel: "The master rewards his servants according to how each has handled his stewardship."

For me to live the mission and purposed life in my new setting, I plan to retire and become actively involved in local, state, and national politics as well as with a variety of civic and social organizations. Fortunately, I have the time to serve on boards, committees, and commissions which are volunteer roles and consequently often are difficult to fill. Being paid is not a factor or an obstacle for me.

My first project already has labeled me as being controversial. My anti-slavery views are not in line with the thinking of many local citizens. Nonetheless, it is a personal mission to which I firmly am committed. Although the institution of slavery is widespread in the area, abolitionary sentiment against it is strong and growing. I am joining their ranks with mind, body, spirit, and money on behalf of liberating the Negro people shamefully treated and disparaged as "colored or darkies."

I have observed the damnable practice of human bondage around the world and find no redeeming value or defense for it. I believe there is only one race of people—the human race—obviously comprised of numerous variances and shades. I am prepared to defend my conviction with scientific and scriptural evidence; but even without it, the worth and dignity of all God's creatures is a birthright sufficient in my value system to support universal freedom from enslavement.

Slaves of General Thomas F. Drayton [139]

The very concept of enslavement is a mystery to me. In addition to my observations during my earliest voyages, I read of its existence throughout recorded history. Even the Code of Hammurabi nearly 4,000 years ago refers to slavery as being natural and commonplace without the slightest degree of shock. Inhumane treatment seems so obviously evil that it would be rejected on its face. To see another creature suffering should be instinctively repulsive. Either from the divine or human point of view, enslavement is damnable in any society—civilized or uncivilized.

Nonetheless, it continues to flourish right before my eyes, condoned or ignored, even in the pulpits of this so-called civilized city. My enlightened Professor Hacker many times drew from Alexander Pope's Essay on Man to explain why choosing bad associates or sur-

"Vice is a monster of so frightful mien
As to be hated needs but to be seen;
Yet seen too oft, familiar with her face,
We first endure, then pity, then embrace."[111]

Alexander Pope

roundings can pollute sound reasoning and corrupt good manners. I recall those lines revealing the three steps that move a person from rejecting an unsavory notion to accepting it—frequent association can bring one to love that which was once thought to be despicable.

Activism

I am attempting to become more diplomatic in my manner; I offer a likely explanation for why it is difficult for me. Twenty years at sea will harden even the most malleable of personalities and dispositions, and I freely admit it has taken a toll on me. I often am characterized as blunt, cantankerous, hard-nosed, plain-spoken, and uncompromising—perhaps accurately so because I usually don't give a damn. A rough-edged temperament can be a handicap, but an asset that brings balance to discussions and decision-making. In other words, I do not go along to get along. Principle, honesty, and the backbone to stand firm against transgressors and transgressions are qualities and traits I admire in people and which I attempt to uphold with my strongest resolve.

I highlight the fact that the greatest leaders in history have tended to project an intimidating countenance—not necessarily intentional—but as an aura that naturally portrays courage and confidence. Think about it; when people select a leader or defender, do they choose one who trembles when passing a graveyard, or when threatened, draws into a shell like a turtle, or plays dead like a possum, or faints like a goat, or runs away like a frightened deer? No, they go for the person who exudes leadership qualities and asserts himself or herself boldly. It is quite simple; follow those who know what they are doing.

"HE WHO KNOWS NOT, AND KNOWS NOT THAT HE KNOWS NOT, IS A FOOL, SHUN HIM; HE WHO KNOWS NOT, AND KNOWS THAT HE KNOWS NOT, IS A CHILD, TEACH HIM. HE WHO KNOWS, AND KNOWS NOT THAT HE KNOWS, IS ASLEEP, WAKE HIM. HE WHO KNOWS, AND KNOWS THAT HE KNOWS, IS WISE, FOLLOW HIM." —PERSIAN PROVERB

Regardless of how I am perceived, not only am I moving full speed ahead on the slavery issue, I have taken on an additional cause in opposition to the Indian Removal Act. Here in 1838, I witnessed thousands of distressed Cherokee Indian families being trailed to the river's edge in Nashville. I saw in the faces of the old and young alike that same forlorn look of heartbreak and despair I remember seeing in the faces of the Pitcairn survivors who had been displaced to Tahiti. Hopelessness paints a cruel portrait.

The Trail of Tears [140]

The first bridge to cross the

164

1853 Suspension Bridge - Nashville [141]

Cumberland River was a three-span covered structure completed in 1823, at the northeast corner of the city square. It was made of wood and iron supported by stone abutments on each bank and two stone piers in the middle of the river. It was the bridge over which I observed the horrible march of thousands of Cherokees crossing into downtown Nashville. It was demolished in 1851 and replaced by a suspension bridge in 1853 tall enough to allow the growing steamboat traffic to clear.

The Indian Removal Act was not outright thievery as such because it was structured and promoted as a fair trade—land and money in Oklahoma for all Indian territory east of the Mississippi River. The controversial Andy Jackson called it "a benevolent policy that brings a happy consummation to the contentious relationship created by the invasion of their homeland by the Europeans."

To me, the entire premise of unwillingly losing one's home proves the trade was unfair regardless of the details. Approximately 15,000 Cherokees alone were a part of the exodus, and 25% of them died en route to Oklahoma Territory. Ultimately, there were five entire tribes removed from their ancestral lands and homes and relocated 2,000

"Stone walls do not a prison make, nor iron bars a cage..."113

—to Althea from Prison
by Richard Lovelace

165

miles away—imprisoned—regardless of how noble it is portrayed. The trade was terrible for the Indians, and the journey was an abominable Trail of Tears.

The Christian missionaries opposed the trade, Davy Crockett opposed it, Daniel Webster opposed it, Henry Clay opposed it, most of the Indians opposed it, and William Driver opposed it in the same spirit and passion which I oppose the institution of slavery.

Subsequently, I encouraged and supported legal action by the Cherokee leaders against the government's unconstitutional action. In 1832, I was thrilled that Chief Justice John Marshall and the Supreme Court overruled the executive branch of our government by recognizing the sovereignty of the Cherokee Nation. However, President Jackson defiantly remarked that "Chief Justice Marshall made his decision, now let him enforce it." Under the Constitution, the President, not the Supreme Court, enforces the court's decisions. President Jackson, in typical fashion, simply ignored the court's ruling."[115]

"Right is right even if no one is doing it; wrong is wrong even if everyone is doing it."[114]

—Saint Augustine

I recognize and appreciate Andrew Jackson's great victory at the Battle of New Orleans. He as an orphan growing up poor deserves credit for

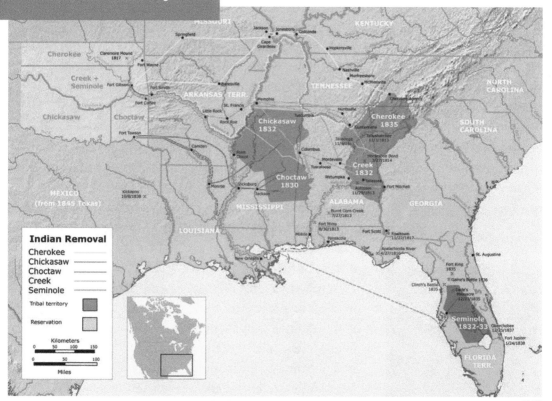

Indian Relocation Paths with emphasis on Northern Route through Nashville [142]

overcoming handicaps and challenges that often thwart commoners in life. He especially was unfairly denied a term as President on a technicality, although he had won both the popular and electoral vote. On balance, perhaps history will judge him as a great president. To me, his role in marginalizing the original inhabitants and rightful landowners of the United States is unpardonable—maybe in time, he will receive their forgiveness, but not ever from Driver!

I could be accused of bias because my wife, the former Sarah Jane Park, is reputed to be kin to Pocahontas and John Rolfe through their son, Thomas, and his daughter, Jane Bolling. Blood indeed is thicker than water, but that association has nothing to do with my feelings that the Indian nations have been dealt one of the worst injustices in history. I pledge to work on their behalf until my last breath.

In addition to the slavery and Indian relocation issues, I plan to spend the next twenty years of my life actively involved in public service roles and activities focusing on social issues and serving my church in various roles. Proudly, I accept the reputation earned as being the opinionated Yankee who is in the middle of anything and everything that is controversial. After all, if matters are not contentious, there is no need for debate or counterposing views to consider. I am pleased to have brought healthy discussion to numerous issues and processes.

Driver on Religion

I am a God-fearing man rescued from the depths of sin and disbelief. As a world-traveler, perhaps I have been exposed directly to as many religious beliefs and practices as any person you will ever meet. Accordingly, it is a subject that has captured my attention and intense study for most of my adult life. I have a selfish motive as well as accountability for guiding a large household in spiritual matters. Hopefully, my paths to religion will provide meaningful insight to others needing direction.

Christ Church Cathedral—Nashville [143]

I served in my church on various committees and as a vestryman leading parish members in the oversight of its operations, its outreach, and doctrinal compliance. I am happy that we were able to resolve a major issue involving accusations of the Rector's autocracy that threatened a split in the membership. Later, I was elected Junior Warden in charge of the upkeep of the parish buildings and grounds.

Although Sarah Jane and I were confirmed at Christ Episcopal, and it is where our children were christened, we did not completely embrace the way Episcopal Doctrine was being interpreted from the pulpit and subsequently applied. In addition, we were strongly displeased with the Rector encouraging members not to vote for Abraham Lincoln, and consequently we dropped out of the fellowship for over a year; I even tried to start a new church called St. James but was denied permission by the Rector; I studied to show myself approved unto God...rightly dividing the word of truth.[116] For my peace of mind and on behalf of my large family, I felt compelled to prayerfully examine thoroughly and explore deeper into this subject with an open mind and heart. I want to feel reassured that we are in the right church home.

Accordingly, I visited other fellowships to broaden my understanding of the beliefs and practices of the various bodies of believers. These included the First Baptist Church, the McKendree Methodist—the largest Methodist Church in the United States, and the First Presbyterian Church, where Andrew Jackson had been a member and James K. Polk was inaugurated as Tennessee's Governor.

As is the case at Christ Episcopal, I confirmed in my examination that all the fellowships were reading from the same book—The Bible—but not deciphering the message similarly. I was struck with the thought of the Tower of Babel account: "let us go down, and there confound their language, that they may not understand one another's speech."[117] God, the creator, designed order—not confusion. There must be a problem either with the document or the interpreters. Since I believe God's Word is infallible, the confusion comes from the heads of those who are confused and not from the inspired text.

The First Baptist Church—two blocks away from Christ Episcopal—was particularly significant to me for two reasons. Foremost, two of my brothers are Baptist preachers, and it was Baptist Doctrine and their influence that led me from a wayward seaman to becoming a true Christian. Secondly, this specific church was engrossed in the middle of the "Second Great Awakening"—a time in our country of renewed spiritual passion and doctrinal examination. Consequently, it had experienced major splintering and divisiveness from forces within prior to my arrival in Nashville..

In a historical context, the "First Great Awakening" took place about one-hundred years ago in Britain and the colonies. It was an age of widespread protestant expansion. Pastor Jonathan Edwards, a Congregationalist minister in Massachusetts, became deeply grieved over the spiritual lethargy and depravity he found in the pulpits, congregations, and

Famous Sermon [144]

communities. His remedy was to preach highly emotional messages of hellfire and damnation which indeed brought the listeners to their knees under condemnation. In his oft-delivered and intense sermon, "Sinners in the Hands of an Angry God," the impact was so frightening that congregations were stirred to groaning, shaking, shrieking, and trembling. In some ways, it paralleled the Shaking Shaker movement in early Colonial America, which was characterized by their highly emotional behavior displayed during worship services.

This Second Great Awakening seems to reflect a more compassionate appeal based on love rather than fear; however, the outward emotional euphoria typical of the first awakening is still apparent. The "Good News" is being preached more widely by gifted evangelists at brush arbor meetings, campgrounds, tent revivals, and the regular meeting-houses throughout the states—particularly in the South and specifically in Nashville.

Campground Meeting

The cause of the earlier referenced discord at the First Baptist Church was Philip S. Fall, the church's second pastor. It had happened before I moved to Nashville. He had embraced a reformation movement within the Second Great Awakening led by Thomas Campbell, Alexander Campbell, Barton Stone, and Walter Scott. Although none were First Church members, their influence through Pastor Fall, persuaded the majority of the membership to adopt the principles and practices of "Campbellism."[1] Five of the original 123 members rejected the redefined fellowship and left to meet in the county courthouse. In 1820 they reconstituted the essence of their former Baptist church. Subsequently, they moved into to a Masonic Lodge where they functioned from 1830 until 1839.

Campground Meeting [145]

The larger remaining group from the original fellowship became the Christian Church and subsequently The Disciples of Christ.[2] They viewed baptism as essential for salvation and were against anything not clearly practiced in the earliest churches. Their movement progres-

[1] The essence of Campbellism, as explained and adopted by the remaining majority, is stated in their following resolution: "That we are unwilling to have our government a creed of man's making. Therefore, we agree to take the word of God alone, as the rule of our faith and practice in all things, and to be governed by its directions so far as we understand them, hereby renouncing all other Creeds, Confessions of Faith, and Rules of Order." The First Baptist Church of Nashville, Tennessee, 1820-1970 p.32 by Lynn E. May, Jr.
[2] Campbell's minister father, Thomas Campbell, defined it as: "Where the Bible speaks, we speak; where the Bible is silent, we are silent." (x) superscript and add citation:

First Baptist Church [146]

sively evolved to what is believed to be a complkete restoration of origional Christianity as established on the day of Penticost—A. D. 33. By 1850, this effort was fully manifasted as the modern day Church of Christ.

First Baptist Church

Concurrent with the splitting of the factions, the Baptists moved out or were chased out of the Spring Street (now Church Street) location to a Masonic Lodge where they functioned from 1830 until 1839 when a new church house was built on Summer Street—my street.

The foregoing account serves as an example of how single denominations and fellowships tend to divide and multiply ultimately into dozens of slightly different bodies and under different names. I observed this kind of evolution numerous times during my exploration and evaluation of religious affairs in Nashville. From my conversations with contacts in other cities, it seems the developments here are but a version of what exists in most every other city across the country.

Summarizing my comparative analysis of Nashville's religious assemblies, bodies, congregations, denominations, fellowships, and societies, it is obvious that all continually are in the process of either being divided, revived or re-formed—becoming more liberal or conservative in the opposite direction from their current positions, but reaching a compromise somewhere between the two extremes.

It is interesting how these developments within the different churches parallel a problem solving methodology that Professor Hacker taught us years ago back in Salem. He asserted that when two or more people, in any domain or context, hold differences of opinion but sincerely seek the truth—not by fighting or quarreling—but amicably by reasoned debate, a classical dialectic method proves to be the most effective. It begins with one of the proponents putting forth a proposition which is called

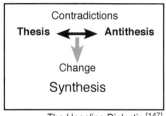
The Hegelian Dialectic [147]

a thesis. The other proponent contradicts that proposition with an antithesis. A synthesis is the desired outcome whereby the two conflicting ideas are reconciled to form a new proposition; in time, the process recycles, and the new proposition is met with a new antithesis, etc. It is amazing how the dialectic method has helped me to tone down and smooth my abrasiveness. I ask more questions before making declarations. I must admit that I thought the technique was original with Professor Hacker, but learned that it similarly was used and taught by Socrates 470-399 B.C.

It seems that many very sincere Christians, by their nature or nurturing, function best in a highly structured worship environment specifically detailed with little leeway. They are most comfortable where the guidelines are unequivocal, the rules are precise and strict, the pathway is narrow, and the directions are presented with carefully chosen words and idioms; conversely, many others are troubled when the line of demarcation is so fine and inflexible that situational discretion, ethics, or objective differences of opinion and interpretation have no voice or latitude.

Regardless, it remains puzzling how two equally intelligent and well-informed people can read a passage of scripture and conclude that it gives opposite messages; each fervently convinced that their translation is the correct one. Since I have been baptized by sprinkling and immersion, I offer an unbiased personal experience as an example.

One evening, I listened to a civil debate on the subject. One proponent argued that according to his church's belief and practice, "immersion" unequivocally was required to complete the process of being "saved." The other proponent countered with the familiar retort about the thief on the cross never being baptized. Then he gave a thought-provoking personal example wherein his brother—a new convert on the way to a "moving stream" to be baptized by immersion—stopped to rescue a mother and child from a coiled poisonous snake; the snake bit his brother who subsequently died before being baptized by immersion. The first debater concluded that the Holy Spirit had provided the baptism and offered scripture supporting his belief. However, the other debater held firmly that the process was incomplete and likewise quoted scripture that tended to support his argument—concluding that for the unfortunate victim, it simply was a missed opportunity and the consequences likely were not good.

The declared loser of the debate subsequently changed his mind and switched positions because of the compelling evidence and persuasive skill of his opponent. He subsequently left his denomination and formed a new fellowship of like-minded believers.

Another notion I found to be profoundly puzzling and objectionable relates to my pet peeve: that being the widely accepted belief that God condones the enslavement of Negroes; even from the pulpits specific Biblical passages are read supposedly as evidence to validate the practice. The jus-

Ham, Japheth, Shem (Sons of Noah) [148]

tification offered is based on punishment for an incident in which they were not involved nor had any knowledge it had taken place.

Accordingly, the account involves Noah and three of his sons, Ham, Japheth, and Shem. Noah, in a drunken state, took off his clothes and fell asleep. Ham happened upon Noah and was so startled he left and told his brothers about what he had seen. They came forth and covered their father's nakedness. For their thoughtfulness, Noah rewarded them, but Ham's descendants beginning with Canaan were cursed forevermore to be servants to the offspring of Japheth and Shem.

Since Ham's descendants are referenced as the forefathers of the African and Middle Eastern nations, the logic suggests they are dark to black in skin tone. As well, the word "Ham" is believed to come from the word Khawm meaning "black, hot, and burnt" in Hebrew. Therefore, a convincing rationale is being preached to the slaves asserting that God wants them to be good and faithful servants, always obedient to their masters, and never to run away. They intentionally are kept ignorant due to the lack of education and are brainwashed to believe God has willed them to be inferior and to live in servitude.[118]

In other words, fair-minded people are asked by many who are insensitive and misguided—not excluding those from some pulpits—to close our eyes and ears to cruel and inhumane treatment being imposed upon an entire segment of God's children. More absurd is to believe even the Creator, and ultimate example of love, fairness, and holiness is complicit in this evil. Although Ham unluckily stumbled upon his drunken and naked father, he was not the one punished for the misstep; it became the fate handed down to the unfortunate victims who were yet to be born—the epitome of God's unlikeness.

I pray for God's judgement on the souls of those who have perverted, mis-

Trail-of-Tears [149]

Slave trader, Sold to Tennessee,

Arise! Arise! and weep no more dry up your tears, we shall part no more, come rose we go to Tennessee, that happy shore, to old Virginia never - never - return.

The Company going to Tennessee from Stanton, Augusta County, the law of Virginia suffered them to go on. I was astonished at the boldness, the carrier Stopped a moment, then Ordered the march, I saw the play it is commonly in this state, with the negro's—in droves Sold.

translated, or mischaracterized Holy Scripture in order to justify the mistreatment of a single human being. The transgressors will get no mercy from me, and I will fight them with all my ability to end this injustice that continues, all around me; so help me, God!

I observed this widespread divisiveness within the church ranks generally to be over subjective non-Biblical matters of man-made doings and which are very much open to debate. I was comforted to hear—almost universally—from every Protestant and Catholic leader, agreement and unity on Christianity's fundamental teachings. They consistently proclaimed that the essence of the Christian Faith is clearly portrayed as the "Good News" found mainly in the four Gospels. I recommend them to all seeking direction and edification in their lives.

Condensed in numerous verses therein, the key requirement is "belief," and the central message is "by grace we are saved through faith, and that not of ourselves; it is the gift of God, not of works, lest anyone should boast."

The most compelling message of guidance I heard throughout the faith community came in a response to a question regarding the requirement to receive eternal life.

> "On one occasion an expert in the law stood up to test Jesus. Teacher, he asked, what must I do to inherit eternal life?... What is written in the Law? He replied. How do you read it? He answered: ...Love the Lord your God with all your heart and with all your soul and with all your strength and with your entire mind; and, love your neighbor as yourself. You have answered correctly, Jesus replied. Do this and you will live."[119]   Luke 10:25-27 NIV

Other Paths

Just as there are over one thousand different distinctions found among Christian bodies of believers, there equally are that many Non-Christian religions worldwide—some are quite large with many followers and are highly developed; some are small and are very primitive. My seafaring career exposed me directly to several of the largest ones, including Islam, Hinduism, Judaism, and Buddhism. As well, I modestly claim special insight into primitive religions because of my many years living in their isolated island environments and unique cultures.

My personal experience led me to Christianity, but I choose not to condemn or exclude the possibility of other paths provided by God. My preacher brothers are quick to chastise me for that point of view. They think I should loudly proclaim that there is only one correct path—which I do believe and embrace. However, I am reluctant to take the judgment seat that rightly belongs to God.

From my earliest days as a Christian, I had a difficult time reconciling my faith with what I observed missing on the numerous remote islands and secluded villages. There were no churches, missionaries, messengers, Bibles, books, or any means that the isolated people could hear the Good News. It seems so unlike a merciful God to condemn and punish the innocent—those given no chance. The entire time I sailed the seven seas and even up until recently, this incongruence continued to trouble me.

Thankfully, as a result of my comparative religion research, I met a gifted theologian while visiting a Presbyterian Bible study. A saintly old lady was leading our study session on the subject of the "unchurched." She provided an answer that had eluded my search all the many years. Amazingly, she could reference and cross-reference every verse in the Bible from Genesis to Revelation. The fact that she was blind was even more startling. The manna I prayerfully was seeking came from her lips as she quoted Romans 1:20 in response to my question.

> For since the creation of the world, God's invisible qualities—his eternal power and divine nature—have been clearly seen "being understood from what has been made, so that people are without excuse."[120]

She confirmed that I was entirely correct that God is not unfair to the innocent or guilty. The unchurched receive their revelations by observing creation all around them regardless of how remote their location. They are enabled to witness the miracles all before them and thereby be judged by whatever light is available. Eureka, I was comforted, perhaps because I was "entertained by an angel unawares."[121]

A frequent response I heard from people who do not attend church services is because there are so many hypocrites in attendance. I presented that reaction to this blind teacher who offered another bead of wisdom. "Since they seem not to mind associating with hypocrites every day in their various walks of life, they shouldn't be bothered being with them in a church service, should they?" Amen.

"Let he who is without sin cast the first stone..."[122]

John 8:7

Christ and the Sinner [150]

My comprehensive study of comparative religions has caused Sarah Jane and me to conclude that Christ Episcopal is a good fit for us. I am at peace for my family to worship there with "the people who are good and the people who are bad...as good or bad as I."—Sam Walter Foss

Politics

Politically, I served my ward as an alderman addressing a variety of issues such as zoning, road maintenance, and crime. Later, I served as a councilman with similar problems but on a city-wide basis. Eventually, I ran for mayor at the urging of several city fathers. Imagine the chances of an out-spoken Damn Yankee abolitionist and the number one antagonist in town becoming the mayor. I was not surprised by losing but was shocked by the amount of support I received. Had my Negro friends and women been allowed to vote, I would have won a term as Nashville's mayor.

The political climate in this city is growing more contentious and divisive by the day; the legality of human bondage continues to fester and clearly is the prime issue. Whigs and Democrats are in combat over which party goes to Washington. Our generally controversial Democrat president, Andrew Jackson, has completed his term and influenced the election of his similarly aligned Vice President, Martin Van Buren. Both are former slave owners as was James K. Polk, our Governor; all were accessories to the Indian removal.

I find Polk's belief in "Manifest Destiny"—that Americans are destined by God to expand its dominion and spread democracy by force to be anything but God-like.[123] The Bible does not promote or condone stealing, including taking by force a person's human dignity. Our nation is in a sad state of affairs, and my very own family is unraveling and rebelling right before my weakening eyes.

American Progress [151]

CHAPTER 12

Driver and Public Education in Nashville

The roots of education are bitter, but the fruit is sweet [124] —Aristotle

By the early 1850s, I became saddened that in my network of neighbors and friends in the Nashville area, many cannot read or write. It is quite common for them to sign a legal document with an **"X."** It has to be affirmed by a literate witness who signs his name by touching the pencil and writing "His Mark" by the X. This witness cannot be a woman or a Negro. Women have no rights to vote, which impacts their equality and the validity of their signature in all matters. Negroes are ineligible because they only are counted as three-fifths of a person; they could not sign anyway because they purposely are denied an education in order to "keep them in their place."

Yes! Old Driver is right in the damn middle of protesting against these continued transgressions on the worth and dignity of any of God's children. First, it was the Indians, then the Negroes, and now—can you believe it—our mothers, wives, and daughters. Routinely, these people are denied equal opportunity based on the circumstance of their birth rather than character. My blood boils to even think about it as my house teems with a full measure of girls. Accordingly, many of the selfish scoundrels in this city find displeasure to see me coming because they know I am quite energized to confront them. They are shocked when I tell them that the gates of hell will be opened and waiting for them—any and all who treat innocent and defenseless people with such disrespect and disregard. Humph! They are either stupid or ignorant—maybe both.

I am continuing to focus on what I believe to be the only solution to these issues and other societal problems; it must come from the hearts and minds of an educated citizenry, and I have found an opening where I might be able to help. I am an admirer of Horace Mann, who happened to be a contemporary of mine back in Massachusetts. He clearly is America's greatest advocate of public education, and for a good reason, I strongly support his mission; either by fate or providence, I recently became involved with his passion—literacy.

In 1852, while in my role as leader of the city's Board of Alderman, the mayor and I commissioned one of our members, Professor Alfred Hume, to visit and study several model educational centers up East. This was to prepare the city's fathers for the implementation of a public school system in Nashville. He specifically focused on the schools in Boston and nearby Salem, which were believed to have the most exemplary educational systems in all of America.[125]

Alfred Hume [152]

Professor Hume returned in August of 1852 to present his report and recommendations to the Mayor and Aldermen. I was there and must tell you that I have heard more than my share of brilliant dissertations, compelling sermons, and passionate speeches but nothing to compare with his report. As well, I have read and studied the most influential masterpieces from the greatest minds and writers that history has given us. I even can remember Professor Hacker's fluent recall of lines about the nature of education taken from Rousseau's "Emile", and his examples of Demosthenes' mastery of oratory as presented in his "Third Philipic," yet with pebbles in his mouth to overcome his impediment. Pericules was another of the gifted with superb eloquence as evidenced in his Funeral Oration, as was John Winthrop's "City on a Hill" that drove the nation to its knees; there are others but none in my lifetime has delivered such a mesmerizing treatise on public education as that we heard that day from Alfred Hume.[1]

Soon after Alfred Hume's report was given, it was adopted as the foundation stone symbolic of Nashville's establishment of public education. Professor Hume died within a year but left his mark and legacy as the father of public education in Nashville, Tennessee. Of all honors bestowed on me in my lifetime, my contribution to this cause as a city alderman is the most satisfying.

The first Board of Education was composed of W. K. Bowling, Allen A. Hall, Alfred Hume, John McEwen, and F.B. Fogg. Under their leadership, Hume

---

[1] There were two thousand copies printed, to which I will provide a link for those interested in reading this history-making report.It will guide others in posterity as required reading for every current educator and those aspiring to be so. It will be gripping for every Board of Education member and higher education department preparing teachers for the future. Lastly, it delivers a compelling message to all citizens desiring to release the chains that bind, discriminate, and enslave God's children. https://archive.org/details/earlyhistoryofna00hume/page/n2

Alfred Hume [153]

High and Grammar Schools opened their doors in 1855 at the corner of Broad and Spruce. It later was combined with the adjacent Fogg School to become Hume-Fogg. In 1856, Hynes School was built on the corner of Line and Summer Street. In 1859 Howard School opened on College Hill.

As the population grew, and the popularity of this phenomenal opportunity increased, so did community involvement and input in terms of the curriculum. Many wanted to include daily Bible readings without sectarian comment along with instructional units addressing industriousness, frugality, temperance, and morality.

The component with religious overtones drew widespread controversy and debate. I attended one of the public forums and received quite a civic's lesson about our Constitution and Bill of Rights—particularly the First Amendment. The establishment clause appeared on a chalkboard as the focus of the discussion: "Congress shall make no law respecting an establishment of religion, or prohibiting the free exercise thereof; or abridging the freedom of speech, or of the press; or the right of the people peaceably to assemble, and to petition the Government for a redress of grievances."

This central proposition related to the separation of religion from government intrusion is a principal reason why our forefathers migrated from the Old World. The protagonists and antagonists referenced James Madison, the author of this amendment, Roger Williams, the founder of the Providence, Rhode Island First Baptist Church—the oldest in America, and President Thomas Jefferson's response to an 1801 appeal from the Danbury Connecticut Baptist Association.

Jefferson responded to the Baptists, "Believing with you that religion is a matter which lies solely between Man [and] his God [and] that he owes account to none other for his faith or his worship and that the legitimate powers of government reach actions only, and not opinions, I contemplate with sovereign reverence that act of the whole American people which declared that their legislature should "make no law respecting an establishment of religion, or prohibiting the free exercise thereof, thus building a wall of separation between Church & State. Adhering to this expression of the supreme will of the nation in behalf of the rights of conscience, I shall see with sincere satisfaction the progress of those sentiments which tend to restore to man all his natural rights, convinced he has no natural right in opposition to his social duties."

Old Driver has "two cents" to contribute to this discussion. It is my belief that our nation was founded mainly by followers and believers in the Christian faith. However, it was not founded as a Christian nation at the exclusion of other faiths or no faith—a major distinction. Throughout history, governments have sanctioned various religions to be theirs officially and often made membership a requirement. Christianity has been no exception; Emperor Theodosius in 380 AD declared that Trinitarian Christianity was the only legitimate religion in the Roman Empire.[126]

Thank God that the early Baptists, James Madison, Thomas Jefferson, and other great minds had the vision to keep secular religion out of the public school classrooms. Those who would prefer Christianity to be taught at taxpayer expense take a shortsighted view; they fail to realize that if it were permitted, paganism and devil-worship would be given a platform as well.

Every benefit of absolute freedom can be abused and produce results contrary to our individual preferences. For example, the debate audience included those on both sides of the Church and State issue. Every time the pro-side speaker rose to give his argument, several antagonists would drown him out by standing, yelling, and stomping their feet. The intimidated moderator asked their leader to restrain his group from interrupting and wait for the con-speakers turn. He snapped back that he and they were exercising their First Amendment right to free speech and peaceably assemble.

As I sat on the front row with the other aldermen observing the reticence of the moderator, I raised my hand hoping for a try at restoring order and was quickly recognized. I stood in silence for a moment, which had a deafening effect on the bully's aggressive manner. I walked slowly and confronted him face to face; he remained standing as the members of his rowdy group quietly began to sink to their seats. I summoned the sheriff to come near us while giving this warning to the agitator; if I am interrupted during what I am about to say, he and I will be handcuffed and taken to jail because a fight surely will erupt. I proceeded with my comments in an atmosphere that had become deathly silent—even a pin dropped could be heard.

"Whoever the Sam Hill you are, I don't give a damn, but hear this: you will be given one more chance to open your disrespectful mouth, and that will be when I am finished talking—not before! Some other fool beside yourself must have explained the First Amendment to you; no sensible and responsible person would ever desecrate and misinterpret its intent and hallowedness the way you and your people have done. You do not have a right to steal another person's time—the one speaking and those listening. You do not have a right to mischaracterize disorderly and unruly behavior as peaceably assembling. Even a child easily could recognize the difference. For your sake, I hope this is a teachable moment. If so, you and your group will sit and be respectful; if not, you can get the hell out of here or be thrown out."

As the saying goes, "fools names and fool's faces often appear in public places;" so other fools likely will show up throughout our lives to disparage and defame the First Amendment. Hopefully, patriots and responsible citizens will be present to call their hand.

The debate resumed in an orderly fashion and concluded with a compromise. Separation of Church and State was never intended to mean that teachers are to eliminate reading selected portions from the Bible or from other books of faith to draw from their history, geography, inspiration, and moral precepts as long as the texts are wholesome and sectarian instruction is avoided. With

Hume Fogg High School [154]

this issue clarified, it is interesting to share what the first weeks were like at Hume Grammar and High School.

Several aldermen and all Board of Education members were there to observe the developments on opening day, February 26, 1855. With no past practice as a pattern to follow, the first days basically were experimental. The Board had hired twelve teachers—one for each grade. All of the Hume Grammar School teachers were high school graduates, and those assigned to the upper grades had completed a minimum of two years of Normal College. The women teachers could not be married, and the girls and boys were taught in separate classrooms. Discipline was maintained with a hickory stick tapped on noggins, standing in the corner, or sitting on the dunce stool for minor violations. The more serious offenses such as laziness, tardiness, and mouthing back at the teacher, earned a visit to the cloakroom to receive the first of two paddlings— one at school and another at home.

It was quite a curiosity to see grown-up farm boys up to twenty-one years of

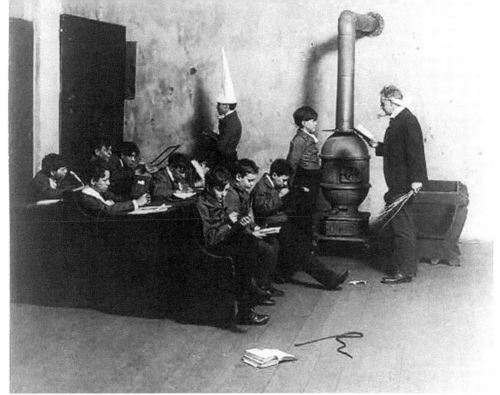

Classroom Management [155]

age reporting to school for their first time—sitting next to 6-year olds—with nothing in common other than neither could read nor write. The staff quickly learned to be creative and innovative—grouping by achievement level rather than by grade or age.

The advanced students mentored those struggling in order for a single teacher to orchestrate a learning experience for fifty or more in a single classroom—a phenomenon indeed—those first schools and their teachers! The other schools during those early years experienced similar growing pains. The basic 3 r's—reading, writing, and arithmetic—necessarily became the priorities and the tools which would enable students to become lifelong learners.

19th Century Classroom [156]

180   *FABLES.*   [PART II.

Fable XLVIII.

The Crow and the Pitcher.

A CROW, ready to die with thirst, flew with joy to a pitcher which he beheld at some distance. When he came, he found water in it indeed, but so near the bottom, that with all his stooping and straining, he was not able to reach it. Then he endeavoured to overturn the pitcher, that so at least he might be able to get a little of it; but his strength was not sufficient for this. At last, seeing some pebbles lie near the place, he cast them one by one into the pitcher; and thus, by degrees, raised the water up to the very brim, and satisfied his thirst.

Morals.

What we cannot compass by force, we may by invention and industry.

The Crow and the Pitcher—Aesop [158]

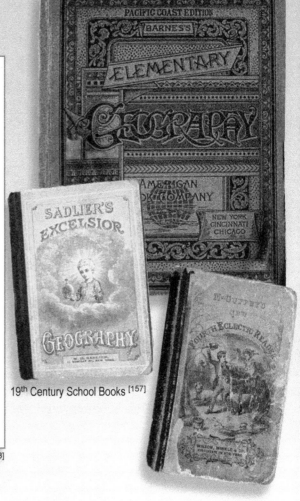

19th Century School Books [157]

The most common early textbooks and reference materials included McGuffey's Readers, Ray's Arithmetic, and Webster's Dictionary. Books such as Aesop's Fables and Grimm's Fairy Tales commonly were used to teach the younger students ethics, decorum, industriousness, frugality, temperance, and morality. McGuffey's graded readers, throughout their instructive stories, served to emphasize virtues and life's values as well.

This marvelous new opportunity that promised to liberate the minds and spirits of our youth was extremely popular throughout the community and especially with the over-aged boys. With tongue in cheek, they convinced their parents how much they loved attending school and how committed to not missing a single day—especially during harvest season.

The Goose & the Golden Egg

There was once a Countryman who possessed the most wonderful Goose you can imagine, for every day when he visited the nest, the Goose had laid a beautiful, glittering, golden egg.

The Countryman took the eggs to market and soon began to get rich. But it was not long before he grew impatient with the Goose because she gave him only a single golden egg a day. He was not getting rich fast enough.

...then one day, after he had finished counting his money, the idea came to him that he could get all the golden eggs at once by killing the Goose and cutting it open. But when the deed was done, not a single golden egg did he find, and his precious Goose was dead.

—MORAL: THOSE WHO HAVE PLENTY WANT MORE AND LOSE ALL THEY HAVE.

The Goose That Laid Golden Eggs—Aesop [156]

It seems that all good things encounter snags and stumble along the way, even toward greatness. This new system of public education was making significant progress and enjoying tremendous support, although not yet ten years old when the clouds above the Republic began to weep—troubled by the most sinister thing imaginable.

Storm Clouds Dividing

Old Glory hanging over the road in the front of my house became offensive and unwelcome to most in the community. We took advantage of the negative feelings to give it a much-needed overhaul; so I pulled it out of the old sea chest, took it into the sewing room, and asked my wife and daughter to repair it. Sarah Jane and Mary Jane not only made the repairs, but they also added a new field of stars along with a white anchor in the lower corner. After dark, I took it to my Union neighbor, Mr. Bailey, and asked his daughters, Mary and Patience, to sew it inside my quilt to make it look like a bed comforter. They graciously did as I asked, and I returned home to conceal it back in the sea-chest. Afterward, there were several searches of the house, but the flag never was found. There was an attempt with a firebomb to burn the house down; however, we promptly extinguished the fire.

On one occasion after Tennessee seceded from the Union, Governor Isham Harris sent a committee to our house and demanded the flag. I met them at the front door and said, "Gentlemen, this is my house, and I am the lord of my castle. If you are looking for stolen property in my house, produce your search warrant." They had nothing else to say and left the premises.

Sometime later an old neighbor, Dick McCann, who was chief of a band of guerillas, came to our door and demanded *Old Glory*. I turned to the squad and said, "Dick, I have known you all your life: if you want my flag, you'll have to take it over my dead body." He turned around and shouted: "Come on boys, let the old man alone."

There is unrest and fracturing in the Union—now in 1858, "…and if a house be divided against itself, that house cannot stand." I hear clearly the cries of certain Southern leaders calling for secession and their flawed rationalization justifying such drastic measures. They attempt to salve their cruelty, immorality, and greed for filthy lucre—the root of all evil.

They cannot fool me, a weather-worn and leather-skinned sailor. At the age of 55, I have witnessed more injustice and suffering than most of them combined. They can't imagine the horror of seeing a living person being dismembered, thrown into a boiling pot and cooked like a chicken; or a child snatched away from his mother with her arms extended, weeping and begging for some merciful soul to help; or the hopeless look of a lonely hollow-eyed black man on an auction-block, stripped to his waist with welt scars on his scrawny back—carved like furrows in an eroded field by a cat o'nine-tails. Damn the bastards who would stoop so low to do this—all of them!

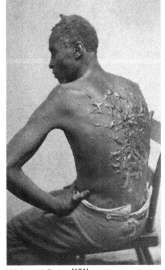

Whipped Peter [161]

It is true that other issues are contributing to the unrest. Certain Northern-imposed trade barriers and protective tariffs are impacting the South unfairly. There are states rights practices whereby the federal government

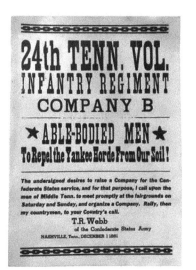

The sale of George [162]

is usurping certain privileges and powers of all states; however, the Southern States, desiring to retain their slave-holding practice, are impacted the most. The United States continued westward expansion also is creating an imbalance between slave and free states.

An example is the passage of the Kansas-Nebraska Act[127] allowing the residents of Kansas to vote on their choice; the region currently is overrun with supporters and opponents equally divided and numbers being killed regularly over the highly contested issue. John Brown, a leading abolitionist, has been instrumental in giving the conflict the infamous nickname, "Bleeding Kansas." He and I think alike except I oppose armed insurrection. It is felt that Kansas eventually will enter the Union as a free state, further unbalancing the legislative equation.

Another factor influencing secession tendencies is a strong sentiment that Abraham Lincoln will win the next presidency and being anti-slavery, his influence and power likely will favor Northern interests.

Recruiting Poster [163]

My opinion is if we take "slavery" out of the discussion, everything else is nothing more than a family feud. If you want a man's allegiance, he generally votes with his purse. I sense that a split is inevitable and soon will happen. This is a pro-slavery city, politically and economically, as evidenced by observation and census records. Even the city government is complicit with its ownership of 26 slaves. In fairness, the vast majority of the citizens never owned a slave, nor would they. It is the powerful few that sustain it, and ending slave ownership clearly will have far-reaching effects—positive and negative. The owners lose, the "freed" will gain, but the assimilation will be costly in terms of education, housing, and jobs. Regardless, it is God's will and must happen.

Lincoln's Defining Speech

On June 16, 1858, at the Illinois Republican Convention in Springfield, Abraham Lincoln initiated his candidacy for the U.S. Senate with his famous "House Divided" speech. It came from the lips of a man whose parents were illiterate and along with himself, had less than 18 months total formal education. It was so politically divisive at the time that it is little wonder why he was defeated.

Conversely, it was so compelling and thought-provoking that it ultimately got him elected as the 16th President of the United States. I am privileged to have met him and have a copy of his most profound message. His words and delivery are evidence of an extremely gifted man called by God to fulfill a noble mission. I will include it with these musings for posterity.

It was a defining moment for Lincoln—his paraphrase of John Winthrop's "City Upon a Hill" message paralleling to the parable of Salt and Light in Jesus's Sermon on the Mount. Likewise, he tells America that she is the light of the world; a city that is set on a hill that cannot be hidden; an expectation that America would differ from the rest of the world and shine as an example to all.

The essential excellence of the message is:

> *"A house divided against itself cannot stand. I believe this government cannot endure, permanently, half slave and half free. I do not expect the Union to be dissolved—I do not expect the house to fall—but I do expect it will cease to be divided. It will become all one thing or all the other. Either the opponents of slavery will arrest the further spread of it, and place it where the public mind shall rest in the belief that it is in the course of ultimate extinction; or its advocates will push it forward, till it shall become lawful in all the States, old as well as new — North as well as South."*[128]

The Tipping Point

Sumter is a fort in Charleston, South Carolina, that is being utilized to defend the East Coast of the United States against foreign invasion. Construction began on it after the War of 1812 and was almost completed when word arrived in Nashville on December 20, 1860, that South Carolina has seceded from the Union, becoming the first Confederate State. Its soldiers, led by Brigadier General Beauregard, ordered the Union Army to vacate. After the Union Major Anderson refused, the Confederates bombarded the area for 34 hours, forcing the surrender. There were two Union soldier deaths caused by an explosion during the celebration of the victory.

Fort Sumter under Confederate Control [164]

Fort Sumter under Union Control [165]

Morse's telegraph was invented in 1844 and was operable in Nashville, delivering the news that the American Civil War officially had begun.[129]

As anticipated, Lincoln's anti-slavery stance got him rejected from the Northern Whig Party led by slavery neutral Daniel Webster. Lincoln and most northern abolitionists united in the newly formed Republican Party. He faced and defeated the slavery supporting Democrat - Steven Douglas, to become the 16th President of the United States. The announcement of Lincoln's victory ignited the secession of other southern states, and by the time of his inauguration in 1861, seven had exited; the Confederate States of America officially were established and soon elected as their president, Jefferson Davis from Fairview, Kentucky. What a coincidence that the divided United States is being led by two conflicted Presidents—both born in Kentucky—125 miles apart. In case any might wonder about William Driver's candidate in that election, I proudly joined with the abolitionist Republicans and voted for Abe Lincoln. I intend to fight the pro-slavery Democrats as long as I live.

Virginia, Arkansas, North Carolina, and Tennessee did not secede until after the Battle of Fort Sumter, which occurred on April 12, 1861. Four additional Border Slave States—Missouri, Kentucky, Maryland, and Delaware did not secede from the Union. Neither did West Virginia which was formed on October 24, 1861, by choosing to break away from the western portion of Confederate Virginia.

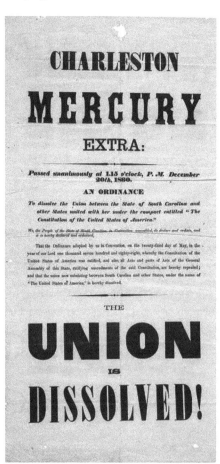

Charleston Mercury on Secession [166]

The Confederate States of America [167]

The following chart shows the order in which the states seceded from the Union.

State	Date of Secession
South Carolina	December 20, 1860
Mississippi	January 9, 1861
Florida	January 10, 1861
Alabama	January 11, 1861
Georgia	January 19, 1861
Louisiana	January 26, 1861
Texas	February 1, 1861
Virginia	April 17, 1861
Arkansas	May 6, 1861
North Carolina	May 20, 1861
Tennessee	June 8, 1861

Following the battle at Fort Sumter, mothers' sons from both sides volunteered by the thousands to join in the fray. Lincoln alone called for 75,000 to help quell the initial rebellion. War plainly has been declared. I am saddened and ashamed to reveal that three of my very own sons, William, Eben, and George, have volunteered to fight for the Confederacy. Thankfully, only a few parents ever will have to relate to the emptiness that comes from being abandoned by their child; it is so unnatural. Conversely, it would be difficult for a child to comprehend why their parent would disown them—equally unnatural. I will share a letter I wrote to my sons, which may help with understanding the enigma. It has to do with integrity and at what point a person of honor and strength would buckle to compromise his principles—his everlasting reputation:

> "My Children, the day I write this record, I am 58 years old. I have lived under the protection of the Flag of my Country unmolested, and nothing could ever make me afraid. For 21 years of my life, I bore that Flag on Shipboard to many lands; and you my Children, William, Eben, and George, have joined this Traitor Band, contrary to your Duty to your Country; Contrary to your Duty to yourself; Contrary to your Duty to God, and contrary to the earnest entreaty of your Old Father.
>
> "In the face of death, I, your Father, record my Solemn Protest against this act of yours. I warn you that a father's blessing can never follow you whilst armed against a Father's Government; and on the Bloody Field when your arm should be strong in a Good Cause, may Terror—the Terror of the Lord God of Israel— make you afraid, for your arm fighting against Him, and Human Liberty, and you, the first of my household, read my eternal resolve to live and die in defense of the Union, the Constitution, and enforcement of the Laws of the United States of America. And when I die, the favor I ask of any of you is to cover my Coffin with that Old Flag, the Stars and Stripes, the Flag of my Fathers!...and (may you) bear me away in silence, without Show! Without a Tear! For unless God stays this Traitor madness which is so abroad in our land, you will require all your Tears for yourselves and Children." Henceforth, we have become strangers in our own family though we share common blood.[130]

One of the benefits of my many years commanding ships is becoming proficient at record-keeping. Making daily entries in the ship's log, journals, and ledgers are requirements that soon became habits and strengths that now are a part of my nature. Consequently, I have been an eye witness to earth-shaking historical developments that are continuing to happen as I write, although some in the future will think it is fiction—brother against brother, father against son, and nation against nation. It is too bizarre to be believable. Indeed it is true, and my hope is that my reporting of the cause and effect will help prevent such occurrences from ever happening again.

Tennessee Volunteers

Tennessee was the last state to secede from the Union. The western part was strong pro-Confederacy, likely influenced by our scoundrel slave-owning Governor, Isham Harris. The east was strongly Union having furnished more volunteers than all other Confederate States combined. I was not surprised because of the great patriotism Tennesseans had shown in previous wars. During the War of 1812, as many as 20,000 volunteered; and in the War with Mexico, the government requested that Tennessee send 2,600 volunteers, but 30,000 reported.[131]

One of my heroes, Davy Crockett, was among those volunteers. He was a Congressman for several terms and opposed Andy Jackson because of his Indian removal shenanigan, but later, he saved Andy's life from an assassin's attempt. I was so proud of Davy and his sharp tongue when he told the voters who rejected him in his last election bid that "they could go to hell, he was going to Texas." He was killed defending the Alamo but was the last man standing—my kind of man!

"I leave this rule for others when I am dead. Be always sure, you are right, then go, ahead."

David Crockett

[168]

As portrayed, Nashville and Middle Tennessee were split between the pros and cons over the slavery issue. My first act of faith and allegiance was to fly *Old Glory* regularly from the rope line we strung across Summer Street. For this, I received praise and ridicule. Many called me an old goat and traitor. I reacted with a stronger choice of words.

My flag remained secretly hidden as I vigorously resisted the demands to hand it over to the Confederates. Even old "Eye Sham" Harris sent soldiers to take it. I told them it would be over my dead body and as well, a good number of them would meet their maker if they tried to physically take it; they left hanging their heads like spineless cowards.

An Unfriendly Wind from the West
—Tis an ill Wind that Blows No Good[132]

Approximately 80 miles west of Nashville in Dover, Tennessee, Ft. Donelson rests on a bluff overlooking the Cumberland River. It is the same water that passed through Nashville a few hours earlier. It actually flows downstream, though westerly, and turns north at Dover—still flowing downstream until it empties into the Ohio River at Smithland, Kentucky. Twelve miles farther to the west on the Ohio River, the Tennessee River joins and the threesome head for a confluence with the Mississippi River at Cairo, Illinois. From there, it connects northward to the headwaters in Minnesota and southward to the Gulf of Mexico at New Orleans—an incredible advantage and convenience. General Grant plans to use this network to pierce the Confederacy and win the Civil War.

No doubt, General Ulysses S. Grant was a military genius and being a West Point man, certainly no dummy. It is reported he drank to excess on occasion—some laughingly suggest for medicinal reasons to treat his migraine

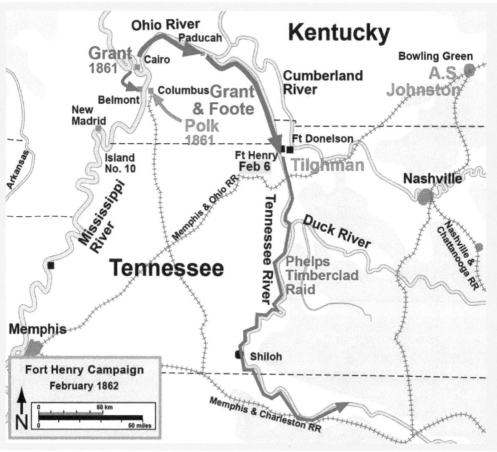

Fort Henry Campaign [169]

headaches. Regardless, President Lincoln probably thought of him as a smarter leader when drunk than most of the Confederates when sober. I cannot speak to that possibility since I faithfully have been a teetotaler since promising my mother I would never drink or smoke.

Nevertheless, Grant believed that by severing the main supply passageways in terms of rivers, roads, and rail tracks, the jugular would be cut—the enemy quickly would be rendered defenseless and vulnerable to land attacks from multiple directions and mechanisms. Consequently, the capture of Fort Henry, which was defending the Confederate's Western Border from water invasion became a high priority. Additionally, a major supply and transportation railway ran nearby all the way to Memphis and northeast to Bowling Green, Kentucky.

The Fort was on the east bank of the Tennessee River guarding a 12-mile wide strip of land spreading eastward to the Cumberland River at Dover, Tennessee. The landmass between the two large rivers extends 40 miles northward into Kentucky, creating a 170,000-acre refuge—the largest inland peninsula in the United States.

The importance of this area to the ultimate defeat of the Confederacy and preservation of the Union cannot be overestimated and merits further analysis and understanding. Although it still is early in this struggle, I believe the War Between the States will be won or lost based on what happens in this specific area. Nashville's communications are connected by overland riders and the telegraph, so I usually am able to receive reports of the activities within days. Though there is enmity between my three of my sons and me, I am writing the developments in my journal as an historical record for my grandchildren to be.

The territory between the rivers served as bounty land for soldiers who served the Union primarily in the Revolutionary War. It was an incentive and payment for their service in the absence of money. The acreage granted depended on the rank and length of service. One hundred acres up to a square mile (640 acres) was not unusual.

The area is overflowing with springs, creeks, churches, schools, and country stores, as well as wildlife. Also, there is an abundance of iron ore, timber, and limestone—the ingredients needed for producing pig iron. I have heard several versions of an intriguing tale, but the one that seems most plausible came from an old-timer who lived his entire life on 1500 acres between the rivers. His version is that a Mr. William Kelly was working at a furnace one-day producing pot metal, a brittle alloy used for quick and easy casting. Once the lava-like concoction reached a "white-hot" point, some boiling bubbles popped out of the mix and stuck to his hammer lying on the ground. When it cooled, he took a rock to knock the residue loose only to find it did not break

Typical Iron Ore Furnace [170]

away as pot metal typically would do. It had become malleable and only flattened when struck. He pondered—pot-metal that can be shaped and formed without shattering, how can that be?

He worked day and night experimenting in search for an answer. His wife and father-in-law thought he had lost his mind and wanted him to see a doctor, but he ignored them, believing it was they who were afflicted and not him. Nevertheless, one day, he noticed that it only was the "white-hot metal" at the surface of the mix that became malleable and concluded it was because the oxygen in the ambient air naturally was mixing with it.

One experiment led to another, and after building a bellows to blow air into the mixture, eureka; he had discovered the process for making steel that was destined to transform the world. Poor fellow; it seems that an Englishman, Henry Bessemer, got wind of the discovery and patented the process in 1855, leaving Kelly out in the cold. However, Bessemer's patent eventually was reversed and awarded to Kelly.[133]

The point of this story is to emphasize that there exists another powerful reason to control the land, natural resources, and the eight iron furnaces operating in the land between the rivers.

The Fall of Fort Henry

On February 6, 1862, it must have been fate that so easily laid Fort Henry in the hands of the Union forces. General Grant was fresh from his victory at Belmont, Missouri. It was across the river from a Confederate stronghold in Columbus, Kentucky. Grant headed south to rendezvous with Admiral Foote, who was underway from Cairo, Illinois toward Fort Henry. His fleet had seven ironclad gunboats and positioned themselves near Panther Creek Island within cannon shot of Fort Henry. The plan was for Foote to attack by water and Grant by land.

The fate I referenced—perhaps the providence of God—is the fact that it rained for many days and nights and completely flooded the grounds of the Fort. General Tilghman realized the inevitable and released most of his men to get a head start on the 12-mile walk to the safety of Fort Donelson in Dover. In time, he lowered the Confederate Flag and hoisted the all-white Flag of surrender. Admiral Foote sent a small boat through the gates of the Fort to pick up Tilghman and return him to the gunboat, *Cincinnati,* for the surrender ceremony.[134]

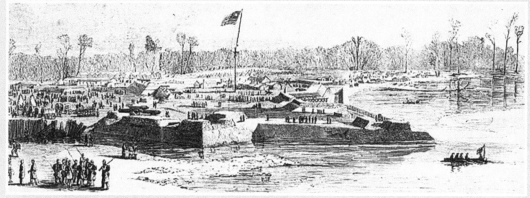
Fort Henry Surrendered to the Union [171]

The battle was over by the time General Grant's troops arrived, so they immediately turned east to pursue the retreating Confederates headed to Fort Donelson. Military strategists forever will wonder what would have happened in this battle if Fort Henry had been built on the bluffs to the north and south instead of the low lying swampy location that foreordained its doom.

In order for the Union gunboats to assist in the attack of Fort Donelson, they had to backtrack north on the Tennessee River to the Ohio near Paducah, Kentucky; then traveling east for 12 miles to the mouth of the Cumberland and 65 miles upstream back to Dover in sight of Fort Donelson. The reception for the advancing gunboats was not as friendly as that at Fort Henry. There were fierce resistance and retreats on water and land by both sides.

The foot soldiers walking from Fort Henry experienced an unusually hot day to travel. The Union men, especially, threw away their coats and blankets, assuming that it was normal since they were far south of their homes in the colder north. However, an unusual cold spell set in that night, accompanied by freezing rain and snow. A report arrived by telegraph that every time the Yanks lit a fire to survive the terrible conditions a sharpshooter would pick them off. It was described as a worse night than Washington's troops spent at Valley Forge.

Ultimately, the combined Union onslaught overwhelmed the Confederates, and General Buckner sent word to General Grant requesting to negotiate terms for surrender. Grant's response was, "No terms except an unconditional and immediate surrender can be accepted." Forevermore Grant will be known as U. S. (Unconditional Surrender) Grant rather than Hiram S Grant.[135]

The heart of the Confederacy had been cut out within days of the surrender. Fort Defiance in Clarksville and Nashville were the next objectives. In preparation, thirteen steamers and two ironclad gunboats were being loaded at the Dover landing with troops from Buell's Army of Ohio. Union General "Bull" Nelson and General Grant were finalizing their plans in the nearby headquarters. Grant believed the next two battles, and especially Nashville would clear

the path to a Union victory. Accordingly, 50,000 Union Troops were waiting to be deployed in the operation facing a projected 20,000 Confederates.

Before departure, General Grant shared a concern that strangely was more troubling than facing an entire division of enemy soldiers. There is a cold-blooded sharpshooter by the name of Jack Hinson, lurking in the hillside somewhere between Dover and Clarksville. He has killed close to a hundred Union Officers on the decks of passing boats. It seems that two of his sons were mistaken as bushwhackers by Federal Troops who killed them, cut off their heads, and stuck them on the gate posts at his home. Although Hinson originally was neutral in the war, he was provoked to swear revenge against the Federal Troops. He had a custom-built rifle capable of accurately hitting targets more than a half-mile away, and it has notches to prove it. There are four Union Regiments looking for him but to avail. General Nelson was warned to keep the troops below deck until past that vulnerable stretch in the river.[136]

The steamers were loaded with about 7,000, men so General Nelson took command of the *Diana* as he bid farewell to General Grant. The gunboats, *Tyler* and *Lexington,* were to lead the flotilla, and the *Carondelet* was already en route. They found Clarksville completely abandoned like a ghost town, so they continued to the Rock City as Nashville was known.

On February 27, 1862, the components merged at the water's edge in Nashville at the city square with thousands of citizens watching from the bluff—without a sound or shot—not even a rebel flag in sight. General Buell had advanced from Bowling Green, Kentucky with about 9,000 troops waiting across the

From Fort Donelson to Clarksville [173]

river in Edgefield. It was an eerie scene—13 steamboats full of troops, three gunboats, and a hospital ship but no resistance. The anticipated 20,000 Confederates were nowhere to be found. Nathan Bedford Forrest provided a rear guard for Hardee's Army of Central Kentucky as it withdrew to North Alabama. A few barking dogs broke the silence as Generals Buell and Nelson went ashore to find the mayor, Richard Cheatham, and several pale-faced aldermen—save one—trailing behind. The mayor formally surrendered the city to the Generals. He was notified that he would be responsible for preserving peace and order through his Provost Marshal, who just happened to be me. I am delighted to report that Nashville has the distinction of being the first Confederate state capital to fall into Union hands. There were large numbers of Union sympathizers who were overjoyed as well.[137]

USS Carondelet arrives in Nashville from Fort Donelson [174]

Hospital ship on the banks of the Cumberland at Nashville [175]

Tennessee State Capital view from Train Station [176]

A New Day in Town

I had been waiting for this moment for months and briskly stepped up after the surrender ceremony to ask which of the two Union Generals was in charge. General Nelson stepped forward in a most imposing manner and answered, "I am in command, who are you?" I paused for a few seconds struck by this man's physical size and assertiveness peering down on me; he was well over six feet tall and at least 300 hundred pounds.

I gathered my wits and answered, "I am a Union man and a retired sea captain who came from Salem, Massachusetts, to live and raise my family here. I will be mighty glad to see that damnable Rebel Flag taken down from the statehouse and have a noble replacement for it." With tears in my eyes, I told him about *Old Glory*, that I had hidden it in my home. If I could have an escort accompany me, I would be happy to fly it on the statehouse. The General responded while giving a friendly slap on my back, "I was a former sea captain as well and believe every good seaman must be a good Union man. I accept your offer and will order an official hats-removed ceremony to raise it to fly over the Capital, night and day, for as long as I am in com-

General William Bull Nelson [177]

Union Soldiers Camped on Capital Grounds in Nashville [178]

mand. Further, I like your blunt-speaking manner and aggressiveness; I will order the mayor to appoint you immediately as the Provost Marshal to help maintain order among the different factions."—a job right down my alley!

One of the first reactions to my new role apparently was the rapid exit from town by the James brothers, Jesse and Frank. They had been living on a farm in the Bordeaux-Whites Creek area of Davidson County under the aliases of Thomas Howard and B. J. Woodson.[138] We were acquaintances but definitely not cordial friends. They justified their practice of stealing as "taking from the rich and giving to the poor"—Robin Hood-type characters. That is not my notion of being charitable, but of further dislike is their pro-slavery sympathy. They are slave owners and strong pro-slavery Democrats. I say it is good riddance.

Nashville began to settle down quickly and returned to its commercial enterprises as though the two-day battle never happened. The war is continuing outside my immediate surroundings, but reports convince me that the Union

Jesse and Frank James [179]

will prevail. I am enjoying my new job as Provost, dispensing orders right and left to the peckerwoods that have been taunting and mocking me.

Riverside Commerce downtown Nashville [180]

Emancipation Proclamation

A momentous event occurred in January of 1863 when President Abraham Lincoln issued a proclamation, a profound declaration delivered by a quill pen, but from the heart and head of a man virtually self-educated. The words and expressions reflect wisdom and eloquence like manna from Heaven formed, fine-tuned, and delivered by a silver-tongued orator. It changed America and the world's view of the worth and dignity of humankind forever. It was an indictment against human bondage declaring "that all persons held as slaves within the rebellious states are, and henceforward shall be free."

I was busy discharging my work as the Provost Marshal when a letter came across the telegraph wires that caught my attention like a bolt of lightning. Underlined were these words written to President Lincoln by the daughter of a slave: "When you are dead and in Heaven, in a thousand years that action of yours will make the Angels sing your praises—I know it." Hannah Johnson

I was so smitten that I copied the entire letter exactly as written. It is the most heart-wrenching, compelling, and thoughtful letter I ever read. When teachers in the future bring the slavery issue before their students, I hope this letter will be a major resource. This humble mother deserves special remembrance and honor for her contribution to influencing history—perhaps a Hannah Johnson medal. She wrote, without corrections:

Buffalo [N.Y.] July 31, 1863, to President Abraham Lincoln

"*Excellent Sir My good friend says I must write to you and she will send it My son went in the 54th regiment. I am a colored woman, and my son was strong and able as any to fight for his country, and the colored people have as much to fight for as any. My father was a Slave and escaped from Louisiana before I was born morn forty years agone I have but poor edication but I never went to schol, but I know just as well as any what is right between man and man. Now I know it is right that a colored man should go and fight for his country, and so ought to a white man. I know that a colored man ought to run no greater risques than a white, his pay is no greater his obligation to fight is the same. So why should not our enemies be compelled to treat him the same, Made to do it. My son fought at Fort Wagoner but thank God he was not taken prisoner, as many were I thought of this thing before I let my boy go but then they said Mr. Lincoln will never let them sell our colored soldiers for slaves, if they do he will get them back quck he will rettallyate and stop it. Now Mr Lincoln dont you think you oght to stop this thing and make them do the same by the colored men they have lived in idleness all their lives on stolen labor and made savages of the colored people, but they now are so furious because they are proving themselves to be men, such as have come away and got some edication. It must not be so. You must put the rebels to work in State prisons to making shoes and things, if they sell our colored soldiers, till they let them all go. And give their wounded the same treatment. it would seem cruel, but their no other way, and a just man must do hard things sometimes, that shew him to be a great man. They tell me some do you will take back the Proclamation, don't do it. When you are dead and in Heaven, in a thousand years that action of yours will make the Angels sing your praises I know it. Ought one man to own another, law for or not, who made the law, surely the poor slave did not. so it is wicked, and a horrible Outrage, there is no sense in it, because a man has lived by robbing all his life and his father before him, should he complain because the stolen things found on him are taken. Robbing the colored people of their labor is but a small part of the robbery their souls are almost taken, they are made bruits of often. You know all about this*

Will you see that the colored men fighting now, are fairly treated. You ought to do this, and do it at once, Not let the thing run along, meet it quickly and manfully, and stop this, mean cowardly cruelty. We poor oppressed ones, appeal to you, and ask fair play. Yours for Christs sake"
— *Hannah Johnson*

Lincoln stood on his principles and by his word. Nearly 200,000 Negroes fought on behalf of the Union, including more than half being former slaves. Because of The Emancipation Proclamation, the economic back of the South was broken. I believe the sins of the fathers will visit upon their children. It is prophetic that the times ahead will be challenging.

COME AND JOIN US BROTHERS.
PUBLISHED BY THE SUPERVISORY COMMITTEE FOR RECRUITING COLORED REGIMENTS
1210 CHESTNUT ST. PHILADELPHIA.

Colored Regiment [181]

The Uncivil War[182]

Appomattox Courthouse[183]

From 1862 to 1865, the fighting raged on. There were thousands of major and minor battles mostly fought in Virginia and Tennessee.

The most prominent battles sequentially were:

1. Battle of Fort Sumter.
2. First Battle of Bull Run.
3. Battle of Shiloh.
4. Battle of Antietam.
5. Second Battle of Bull Run.
6. Battle of Chancellorsville.
7. Battle of Gettysburg.
8. Siege of Vicksburg.
9. Battle of Atlanta
10. Battle of Appomattox Court House[139]*

*The war ended on April 9, 1865. General Robert E. Lee surrendered the last major Confederate army to General Ulysses S. Grant at Appomattox Courthouse, Virginia. Over 100,000 citizens, 360,000 Union soldiers and 260,000 Confederate soldiers died. One of the latter was my son. Thousands more were wounded and maimed for life.

...in the beauty of the lilies Christ was born across the sea, with a glory in His Bosom that transfigures you and me. As He died to make men holy, let us die to make men free...God's truth is marching on.[140]
—The Battle Hymn of the Republic

Lee's Surrender[184]

CHAPTER **14**
Nashville Post War

Ward Lamon was a personal friend, confidant, and bodyguard of President Abraham Lincoln. On April 12, 1865, the President shared with him a strange dream he described as seeing himself entering a room in the White House where people were sobbing at the side of a covered corpse. He asked who was dead, and a guard said: "it is the President; he was killed by an assassin."[141]

Three days later, the 16th President of The United States was assassinated by actor John Wilkes Booth at a play in Ford's Theater in Washington, D.C. Booth was a Confederate sympathizer strongly opposed to the abolishment of slavery.

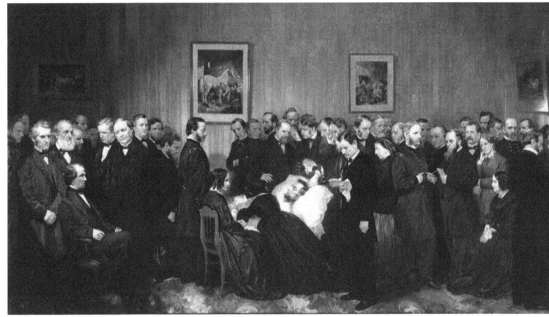

Lincoln's Final Hours [186]

In my view, Lincoln died a martyr's death because he was the catalyst for the abolition of what Booth loved more than life. Abe had a premonition that it would happen, and like Socrates who drank the poison hemlock, he would not compromise his equality principles even to save his life…

"…we hold these truths to be self-evident, that all men are created equal, that they are endowed by their Creator with certain unalienable rights, that among these are Life, Liberty and the pursuit of Happiness" —The Declaration of Independence

Lincoln's Funeral Train, The Old Nashville [187]

I have met this humble man and am deeply saddened by his passing, but I realize that it had to end this way. I doubt that in a hundred years another will come along with his greatness; and a word to his successor, Andrew Johnson:

Johnson and Lincoln-Reconstruction Era [188]

just remember that you never want to follow a legend in life in any capacity. You will be unable to be his equal and already are doomed if you try. I know Andrew Johnson also and believe the future does not bode well for him. He would have been better off by remaining in the tailoring business.

Reconstruction -The Conflict Continues

To facilitate the integration of Negroes into a free society, Lincoln envisioned establishing a Freedman's Bureau in March of 1865. In a hard-luck defense of Andrew Johnson, he inherited an unmitigated mess trying to piece together three factions—the four million freed slaves, the disenchanted Rebels, and the victorious Unionists. President Johnson continued with Lincoln's plans for Reconstruction focusing on uniting all of the States and forgiving the citizens of the South for seceding if they pledged an oath of allegiance to the Union.

Johnson's Reconstruction [189]

The goal to unite a fractured nation into one happy and loving family living in peace and harmony is proving to be a colossal task. Obviously, most of the leaders have never been involved in such an undertaking. They are finding the seven deadly sins not being cooperative with such a lofty mission.

Accordingly, there is considerable pushback. It began with the assassination of our President. It continues with insurgencies such as the Ku Klux Klan—The Invisible Empire—with its roots in Pulaski, Tennessee.[142] Confederate General Nathan Bedford Forrest was the first Grand Wizard and leader of the National Association which often meets in Nashville. The Klan uses fear, intimidation, and sometimes violence to thwart or deny the rights of the freed men and women. Also, they discriminate against Catholics, Jews, and the anti-slavery radical Republicans. Cross-burnings and even lynching are commonplace. There are other secret societies known as vigilantes and night riders similarly wearing hoods and using the same terrorist tactics.

Now coming into play is Newton's third law of motion that professor Hacker taught us: "for every action, there is an equal but opposite reaction." A counterforce in the name of William Gannaway "Parson" Brownlow entered the scene with the reputation of being the most controversial person in the history of Tennessee. I am somewhat disappointed that he is not considered to be the second most controversial.

Nevertheless, he was orphaned at the age of 10 and lived with an uncle who taught him the carpentry trade. After attending a revival at age 21, he "got religion" and became a circuit-riding Methodist preacher. Being strongly disposed to speaking his mind, he continually was involved in quarrels with other denominations and particularly the Baptists. He was sued for libel by one such member and paid a $5 fine. That got him kicked out of the North Carolina associations all the way to Maryville, Tennessee.

Soon, Brownlow became crossed with the Presbyterians and was transferred by the Conference Bishop to South Carolina—again into a hotbed of Baptists. There, his continuing contempt and disparagement of them and the slavery-supporting Democrats, got him chased out of South Carolina back to Tennessee to avoid being lynched.

In addition to his fiery personality, I admired Parson Brownlow because he was a eloquent orator, writer, and savvy politician. He started a Republican-slanted fiery newspaper in East Tennessee, which created some bitter exchanges with a Democrat-leaning newsprint editor. One of his editorials precipitated a street fight in Jonesboro with Democrat Landon Hayes. The Parson was battering Hayes on his head with a sword cane when Hayes drew a gun and shot Parson in the leg. The fight ended, but they clashed continually for many years.

[190]

Parson moved his newspaper to Knoxville, where again, he was struck on the head by an unknown assailant and hospitalized. Subsequently, he was arrested by confederate authorities and charged for treason.

He despised Andrew Johnson and called Andrew Jackson a curse on the nation. Oddly, he became extremely popular as a lecturer and writer, which ultimately gained him the governorship of the State of Tennessee in 1865. As governor, he enabled Tennessee to be the first Confederate State to be reunited in the Union—a coincidence since Tennessee was the last state to secede.

Brownlow pushed hard to give freed slaves their rights. As well, he took on the Ku Klux Klan and Nathan Bedford Forrest by suggesting it was entirely proper for Klan members to be shot on sight. I am two years older than the Parson, but we definitely are "birds of a feather" on most subjects and opinions, es-

pecially being friends of the Negroes. He learned to despise the Nashville slavery-supporting Democrats calling them dunghills, and my feeling is mutual. Parson Brownlow is a true champion for the downtrodden and a spokesman for the timid. He was a lifelong Methodist, a staunch Republican, and an unwavering abolitionist. I liked him as governor!

Reconstruction essentially failed in its objective to bring the South into a harmonious relationship with the North, and to successfully integrate the freed slaves into society. It seems that crooked politicians can always find a way to skirt the law to accomplish their selfish wishes—at

William Gannaway Brownlow [191]

least temporarily. In this case, their ploy is personified in a white Yankee actor from New York named Thomas "Jim Crow" Rice.[143]

Thomas would blacken his face, put on ragged clothes, and go on stage lam-

pooning Negroes as dim-witted, illiterate, lazy, and superstitious "jolly dancing niggers"—his words. Certain perverted fools find that demeaning and insulting fellow human beings is quite funny and entertaining. However, in so doing, they expose their motive for diminishing and devaluing the worthiness of the Negro in his quest to participate in the community on the same level as the white man. I intentionally did not include "the same as the white woman" because, in many respects, they are treated essentially the same.

The separatists deceitfully found a way to limit the inclusion of people of color. It is somewhat like a June bug on a string[144] being free to fly, yet its range remains in control of the one who holds the string. Similarly, human bondage

Jim Crow Minstrel Dancer [192]

advocates and segregationists found a less offensive name to shroud their evil intent.

The parallel I offer is that "Jim Crow" is a mirage, a name that misleads one to believe something not true yet portrayed as authentic. The unwilling separatists created an illusion known as Black Codes—simple prohibitions but with powerful implications. Collectively, they plainly meant: "not for the colored." Actually there were laws passed by the Southern States at the end of the war, condensed and posted on signs with the intent and the effect of restricting the freedom of Negroes.

Freedmen's Segregated School [193]

Colored drinking fountain from mid-20th century [194]

The gist is that the slave is now freed, but not free to drink from the white only water fountain, or use the white-only rest room, or live in the white-only neighborhood, or go to the white children's schools, or be paid the same wage for the same job. The "colored" woman can come in the back door and cook the meal but cannot come in the front door and sit with the white folks or eat with them. Indeed, Jim Crow laws are fully implemented in Nashville and accomplishing exactly as intended.

"Colored" cabins [195]

No, my colored friends, you are not truly free, and it will be many decades—maybe even a century—before you are; but you will be. The Abe Lincolns, John Browns, Parson Brownlows, Booker T. Washingtons, Dred Scotts, Sojourner Truths, Frederick Douglasses, Harriet Tubmans, Hannah Johnsons and William Drivers of the world will see to it, so help us, God!

Post Reconstruction

In 1868, the unpopular Andrew Johnson, was not chosen by the Democrats to run for reelection. The party chose Horatio Seymour, the Governor of New York and a white supremacist. The Republicans selected the Union hero U.S. Grant, who won easily. The election stirred up some harsh feelings and animosity that Lincoln had hoped would be buried forever. The separatists were likened to dogs for wanting to dig up the bones of hatred and continue the conflict.

Lincoln - Douglass [196]

"Though the mills of God grind slowly, yet they grind exceeding small;
Though with patience He stands waiting, with exactness grinds He all."[145]
—Henry Wadsworth Longfellow

Trouble in Bluff City

Although retired to volunteer public service roles, I was kept very busy with

Peace or War [197]

tasks at my church and with the local government as an alderman, council-man, and Provost Marshal. As well, I have been called to extra duty at the hospitals, helping with the overload of the sick and infirmed. As Provost, I have encountered the biggest challenge of my life. It is appalling, because of the cause, that I report the number of infectious disease patients multiplying so rapidly that our hospitals and infirmaries are unable to provide adequate care.

You will recall that because of our strategic location as a center for transpor-tation, trade, and manufacturing, General Thomas selected Nashville as the Union's Operation Headquarters; this happened concurrently with the city be-coming the first Confederate State Capitol to fall. Consequently, disproportion-ate numbers of men—well in the high thousands—are patronizing or living in the city and surrounding areas. The obvious reason relates to nature's drive for intimacy; this requires no explanation to the mature other than it is innate and necessary for the propagation of the species. I have observed this interaction play out in societies around the world in a consistent and predictable pattern, occurring endlessly and naturally without alarm or concern.

There are codes of conduct, doctrines, morals, principles, religious precepts, societal standards, and prevailing practices that influence and govern the out-comes of these kinds of liaisons. They are for the churches, homes, and schools to address, but our immediate concern is not the morality, rightness, or wrong-ness of the behaviors; rather, it is the reality of the fact that a sexual-related crisis is overwhelming the city. Promiscuity and "pay for play" appear to be the catalysts that must be addressed.

The Aldermen and City Council, at the behest of the Mayor, formed a com-mittee including a cross-section of citizens to study the matter and make rec-ommendations for corrective action. Our findings included that:

1. The hospitals are overflowing with venereal disease patients.

2. At least one in every ten soldiers is reported to have contracted gonorrhea or syphilis, not counting the other types of infections. The rates are believed to be even higher for troops garrisoned in the city and those nearby.

3. The area around the Capitol Building and Smokey Row has given Nash-ville the infamous reputation as being a colossal red-light district inundated with numerous brothels and other houses of ill-repute operating around the clock. There were 207 acknowledged prostitutes, working out of the area ac-cording to the 1860 census. Within three years, the number had grown close to 2,000, and by now, we have lost count.

4. The lower part of town near the river is considered to be a public disgrace and nuisance judged to be off-limits for respectable people. There are "street women and ladies of the night" on every corner marketing their wares along with "pimps" hustling and hawking on their behalf for commissions.

5. One of every three post-war deaths is due to venereal disease, for which there are no effective treatments. More soldiers are dying from diseases than from bullets.

6. The primary reason given by women for engaging in this so-called "world's oldest profession" is economic necessity, not sexual gratification. Women undeniably are victims of discrimination—without voice, often in church, their homes, the ballot box, and at the pay station. Many are unable to find employment elsewhere and are the only breadwinners for their households. This enterprise provides a lucrative income to pay for housing, food, and daycare for their children; unfortunately, others are working to pay for their drug and alcohol addiction.

Some soldiers misguidedly gave themselves syphilis intentionally, trying to self-immunize to avoid the deadly smallpox. They are cutting themselves and rub fluid from another's infected pustule sore. Trying arm-to-arm vaccination seems to them like the best, if not only, solution. Ironically, there is no cure for syphilis either.

Our committee held several meetings progressing from the sublime to the ridiculous. It became quite comical to hear some of the proposals offered to alleviate the VD problem. Respectfully speaking, a saintly and quite elderly "old maid" suggested that we initiate a campaign urging abstinence. Accordingly, she reasoned that through the messaging and guidance of our teachers, preachers, and community leaders, if we stopped the cavorting and frolicking between the sexes, we would eliminate the problem at its source—"out of sight—out of mind." There is a degree of truth to her proposition—if it can be enforced.

Smallpox [198]

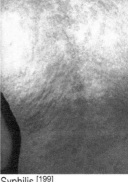
Syphilis [199]

Another notion put forth is that we arrest and impose fines on people who come down with venereal disease; obviously, they did not exercise proper precaution. Indeed, jailing folks would stop the liaisons as long as they remain incarcerated—another solution indeed, but not practical.

Then we heard a bead of wisdom from a most unlikely source, a cattle farmer. His approach was applying common sense as revealed through a parallel to farm life.

"Committee members, just last week, it was time for one of my heifers to 'come in heat.' In case any of you are unfamiliar, that happens first at about

15 months of age and thereafter every 21 days. Nonetheless, she had been limping around a few days for some unknown reason, so I separated her behind a barbed wire fence to another field away from my bull so I could keep a closer eye. I pretty much forgot about monitoring the situation for a few days until—while working in the barn loft—I heard an uncanny commotion accompanied by clanging and bellowing like an elk horn along with a tornado passing through. I rushed to see what was hap-

Nashville's Hells Half Acre – Slums, Saloons, and Brothels [200]

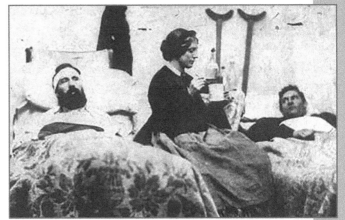

Union nurse Annie Bell in Nashville [201]

pening. I must admit to a chuckle and head shake because Mabel had 'come in heat,' and Samson had torn down an entire section of fence to get to her. By the time I got closer to inspect, he had barbwire scratches and was bleeding all over from cuts. Apparently, he had answered nature's call, and both had returned peacefully to grazing as though nothing had happened.

"Folks, I see nature revealed daily with roosters chasing the hens, the ram and ewes, the tom and the mollies, and the sires and the dames as they instinc-

Brothel [202]

tively—without lessons— fulfill their roles God gave them instinctively in his marvelous creation. I do respect the other solutions offered and those who gave them, but they suggest that we try to modify nature. My farm life

Hospital for Infected Prostitutes in Nashville [203]

has taught me a valuable lesson that applies to everything in life; if you try to combat nature, you will lose the battle—plain and simple.

"The solution lies in the treating and curing the infection causing the disease. The first step has been taken by General Rosecrans with his order to legalize prostitution in Nashville. With this provision, conditions and standards can be enforced to control it."

"The meeting adjourned with the committee members being dumbfounded by the homespun wisdom. On August 20, 1863, Nashville, Tennessee became the first city in the United States to legalize prostitution."

Accordingly, each prostitute was required to purchase a license from the Provost Marshal to practice her profession. She must be examined weekly by a certified medical doctor who will confirm her health fitness. She will pay a weekly tax of fifty cents to defray the cost of operating a new hospital specifically designated to address the treatment and care of VD patients. Failure to practice in the trade without a license and health certificate will land the transgressor in the jail workhouse for 30 days.[146]

In the summer of 1864, Memphis, Tennessee, followed Nashville's lead and legalized

Provost Marshal granted license to Prostitutes [204]

prostitution in its city as well. The concept and program was a great success, and in time, the transmission of venereal diseases was virtually eradicated at the source—the professional madams in the sex trade industry. The program and practice generated quite a number of sermons from the pulpits of the city, which in turn produced a good deal of squirming by some of the brethren in the pews.

A hospital for Federal Officers [205]

Reconstruction's Final Act

During the period of time from the surrender of Nashville in February 1862, through the end of reconstruction in 1877, the South remained under the control of federal military forces. During that 15-year stretch, transitioning into one united country was challenging. Of greatest significance was the passage of three so-called Civil War Amendments, which addressed the emancipated slaves and were the first amendments to the U.S. Constitution made in 60 years. They are the 13th passed in 1865, which abolished slavery and involuntary servitude, the 14th passed in 1868, which addressed citizenship rights and equal protection under the law, and the 15th passed in 1870, granting Negro men, not women—black or white—the right to vote.[147]

Contrary to the interests of these amendments advancing the Negros' cause and rights under the Constitution was the strangest and most disputed election in American History. Tennessee and Nashville, in particular, became a major focal point of this development. In the presidential election, Samuel J. Tilden of New York won the popular vote against Ohio's Rutherford B. Hayes and was ahead in the electoral tally by 184 to 165. However, due to some old-fashioned political horse-trading, a compromise occured, shifting 30 electoral votes to Hayes along with the presidency of the United States.

The curious part of this so-called Compromise of 1877 was Hayes agreeing that the United States federal government would pull out the last troops in the South, formally ending the Reconstruction Era. Some erroneously suggest that Nashville was awarded a Federal Customs House as a part of the deal. However, it was Florida, Louisiana, South Carolina, and Oregon, being involved in the disputed electoral votes and not Tennessee.

It remains a mystery to others why Nashville even qualified to have a Customs House. Traditionally, the purpose is for the oversight of importing and exporting goods into and out of the country, and as such, they naturally are located at a seaport or river with access to an ocean.

It is a real stretch to say that Nashville met the typical qualifications. However, the city technically does have access to an ocean via the Cumberland, Ohio, and Mississippi Rivers flowing into the Gulf of Mexico at New Orleans.

Nevertheless, the cornerstone of Nashville's Customs House was laid by the 19th president, President Rutherford B. Hayes, in 1877, on the first visit of a president to the South since the Civil War. [148] Thereafter, Recontruction ended in the South with the removal of federal troops, and Nashville—a most unusual seaport city—strangely got itself a Customs House.

As usual, the unfortunate Negroes found themselves back in the same predicament as before the passage of the Civil War Amendments. However, with tongue in cheek, I am sure they are comforted to know their plight no longer was enforced "by law" (de jure); their discriminatory treatment became "in fact" (de facto). Most everything is the same as before except the chains are missing, as are the overt prohibiting signs.

The separatists again found a loophole and this time, I am sorry to say, they colluded with a Republican President to pull it off. It is clear that time and additional legal action will be necessary to correct the unfairness of this setback. My experience has taught me that social change of this magnitude takes three generations to be achieved and for assimilation to occur. The wheels of justice necessarily will grind slowly but exceedingly fine. The great-grandchildren of my Negro friends can expect unmitigated impartiality genuine equality of opportunity to occur—God willing.

Customs House in Nashville, Tennessee [206]

CHAPTER 15

Driver's Final Years

Captain William Driver [207]

The last two decades have been the busiest for me as any in my lifetime. I have served in several appointed and elected offices such as that of alderman, councilman, Recorder of Claims, Tax Assessor, Chairman of the Citizens Relief Committee, and I even ran for mayor. I have volunteered to serve on numerous commissions, councils, boards, and committees, including several as the leader and spokesman. My roles have extended to levels of engagement with other aldermen, councilmen, mayors, congressmen, senators, gover-

Farm Theft [208]

Buena Vista Ferry Request [209]

nors, and three presidents. As well, I continue to be a community activist focusing on social, civic, health, and religious matters while interfacing with all races, colors, and creeds.

One of the more satisfying roles was that of serving on the Commission of Claims as the Recorder. The occupation

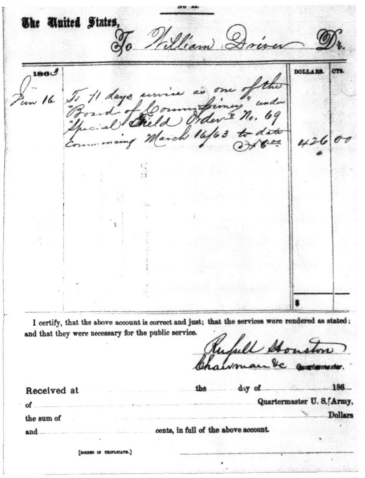

Driver's Pay [210]

of thousands of Union armed forces in Nashville during and after the war inflicted considerable property damage to schools, churches, hospitals, and livestock of the citizens. Damage especially was to the downtown churches with the exception of Christ Church—my home church. The commission's job was to assess the validity of claims of damage and determine the appropriate reimbursement. This work is very fulfilling because it is helping innocent people in dire straits recover from the ravages of war.

Although my jobs had a worthy mission and purpose and are personally rewarding, the pace is finally catching up with me. The salt is losing its savor. I am growing tired and weary now in my eighties, already having out-lived all of my forefathers, several siblings, children, and most of my friends. I am well beyond my allotment of three score and ten, so I face and accept reality. My faith makes me unafraid, and in fact, I am an obedient, willing, and ready servant prepared to cooperate with the inevitable without flinching.

As a world traveler to the four corners of the earth, I have seen and done most everything the hungriest of hearts and minds could desire or that coins could purchase. Queens, kings, emperors, and rulers of empires have nothing at their beck and call that has been out of the reach of William Driver. All of the joys and sorrow of life have touched my emotions, and some have cut out a piece of my heart. Losing children, mates, brothers, and sisters are the heaviest burdens, yet life has a way of counterbalancing the good and bad. The ecstasy that follows success and the agony that accompanies failure are well known to me.

The deeper that sorrow carves into your being, the more joy you can contain...some of you say, "Joy is greater than sorrow," and others say, "Nay, sorrow is the greater." But I say unto you, they are inseparable...When you are joyous, look into your heart and you shall find it is only that which has given you sorrow that is giving you joy. When you are sorrowful look again in your heart, and you shall see that in truth you are weeping for that which has been your delight.[149]

—The Prophet by Khalil Gibran

Lingering Thought

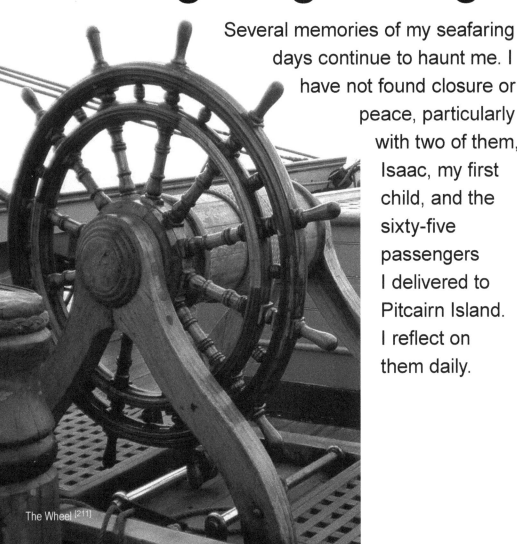

Several memories of my seafaring days continue to haunt me. I have not found closure or peace, particularly with two of them, Isaac, my first child, and the sixty-five passengers I delivered to Pitcairn Island. I reflect on them daily.

The Wheel [211]

Infectious disease epidemics in the south sea islands during that time frame were widespread and devastating. The lower than normal life expectancy was such that Isaac, who now would be in his mid-sixties, likely is not alive. Regrettably, it has been over fifty years since we have seen or heard from each other. He seemed unusually bright and ambitious as a child. I pray that he was happy and achieved well.

My unconventional island marriage to his mother, Talei, was never sanctioned by me or viewed as authentic. Regardless, the circumstances did not provide options for me to live on the island or for them to join me in the United States. Talei had other husbands, which was the common practice in her polygamous culture. That practice alone placed my separation from her in an entirely different perspective. In any context, it was the unknown about Isaac that has left me discontented. If he lived a full life within ordinary circumstances, surely there are thousands of Driver cousins dispersed around the world today. I can only hope that in time, they will cross paths with their cousins from my other marriages.

Pitcairn and Norfolk

Unlike entirely losing contact with Isaac, I have been able to follow many of the changes linked to my acquired friends who were the family members of the mutineers. The phenomenal increase and distribution of the nation's newspapers during the mid-1800s and the popularity of the Pitcairn story combined to attract a large following of readers. Advancing age and a degree of melancholy have nudged me to become a prolific letter writer, mainly compiling memoirs for my children and grandchildren. It should be of no surprise that my direct involvement in the story has motivated me to continue to follow and write about the developments as well.

You will recall my earlier account of the meeting with Queen Pomare IV in Tahiti and my agreement to transport the mutineer descendants back to Pitcairn Island. Sadly, though decades have passed, I remain without a word from any of those who were befriended by my crew and me. Nevertheless, it is a significant point to note that the last authentic mutineer, John Adams, died while living on Pitcairn Island in 1829.

The following account updates previous letters to keep my journal current and my grandchildren aware of the latest developments; or whoever may find it of interest. I assume you recall the events related to the famous April 28, 1789, mutiny on the *HMAV* [Her Majesty's Armed Vessel] *Bounty*—a merchant ship purchased by the Royal Navy for a botanical mission involving breadfruit trees. Had it not been for this mission and the subsequent mutiny, I never would have been involved in this event that changed the direction of my life.

Western view of Pitcairn Island [212]

Captain Driver's Final Pitcairn Chapter

"I shall never forget August 14, 1831, when the Charles Doggett departed Papeete, Tahiti en route to the passengers' former homes on Pitcairn Island. We delivered them to their familiar landing in Bounty Bay on September 3, 1831.

...1767, "IT IS SO HIGH THAT WE SAW IT AT A DISTANCE OF MORE THAN FIFTEEN LEAGUES [THREE STATUTE MILES], AND IT HAVING BEEN DISCOVERED BY A YOUNG GENTLEMAN, SON TO MAJOR PITCAIRN OF THE MARINES, WE CALLED IT PITCAIRN'S ISLAND."

"In a flashback, their forefather mutineers forcibly took command of the *Bounty* and set Captain Bligh and 18 of his loyalist crew, adrift in a utility lifeboat. Eventually, they found their way back to England, where the somewhat contentious Bligh was exonerated of any wrongdoing and lived to gain retribution.

"The leader of the mutineers, First Mate Fletcher Christian, and the remaining crew attempted to hide on the secluded island of Tubuai located halfway between Australia and South America. However, due to conflict with the natives, they returned to Tahiti, where 14 of the mutineers and two of Bligh's loyalists decided to stay. Christian continued to sail with the remaining nine mutineers, six Tahitian men, eleven Tahitian women.[1] They accidentally came upon and landed at Pitcairn Island on the 23rd of January, 1790. It is a volcanic island in the South Pacific approximately 1,400 miles southeast of Tahiti. The mutineers found it to be uninhabited but discovered rock-wall etchings and crude tool fragments identified to be of early Polynesian origin.

[1.] The number and composition of persons coming, going, and staying at Tahiti, Tubuai, Norfolk, and Pitcairn vary slightly depending on the sources referenced. This is explained in part due to births and deaths within the respective time frames. The authors chose the most frequently quoted numbers for this account.

"There were no plans ever to leave the island, so in fear of being discovered, the leaders decided they should destroy the Bounty. First, they gutted the ship by removing all the tools, utensils, and wood for shack building. Thereafter, they unloaded the remaining cargo, which included a variety of livestock, breadfruit plants, and all kinds of seeds. They had everything needed to begin a new settlement but drinking water—a significant challenge. Except for a few small freshwater streams, the primary source was rainwater which they caught and stored in barrels. After the ship was but a shell, it was burned and sunk in what became known as Bounty Bay.

"The settlers were prepared to farm on the fertile volcanic soil and grow food staples such as arrowroot, beans, breadfruit, cabbage, sweet potatoes, tomatoes, bananas, citrus fruits, melons, and pineapples. Coconut trees already were plentiful as were fish easily caught along the shoreline. The transported goats thrived throughout the hillsides and provided an abundance of cheese, milk, and meat.

"Initially, they found Pitcairn Island to be a paradise and an unmatched hiding place. In time, they learned that their new location was one of the most isolated spots in the world. All cooperated effectively in blending their unique skills and labor. Each was expected to work, providing for themselves and the settlement in general. A loosely structured form of government was established, including standards and enforcement procedures regarding civility, law, and order.

"Soon, children were being born. The first was a son to Isabella and Fletcher Christian whom they gave a strange but logical name, Thursday October Christian. Early on, the environment seemed to be utopian, but the tranquil setting was soon to change—it was the calm before the inevitable storm. As history has proven, humanity left to its own natural devices, predictably will destroy itself. The nature of human nature began to unfold. Racial tension developed between the natives brought from Tahiti and the Europeans. Jealousy and enmity erupted over the imbalanced partnerships—15 men and 11 women.

"McCoy, a very resourceful mutineer, discovered how to distill brandy from ti-root and soon built a very productive still. He and several others, including some of the women, continually were drunk from the concoction. Soon, in addition to accidents and diseases, the population further was impacted negatively by alcoholism becoming pervasive as was promiscuity, infidelity, rape, suicide, and murder. Even the creative McCoy tied a rope from his neck to a large rock and jumped off a cliff to his death.

"From 1790 to 1808, the colony progressively declined into a state of degradation and iniquity. Most of the mutineers had been murdered except for John Adams, Edward "Ned" Young, and Matthew Quintal. The so-called crowning

The grave of John Adams on Pitcairn Island [213]

blow came in 1799 when Adams and Young got Quintal drunk and killed him with a hatchet.

"John Adams, also known as Alexander, was the spiritual leader and teacher for the children.[1] He described the view and status of the women as being deplorable. Generally, they were abused and considered to be property the same as animals. They routinely were passed around from man to man for sexual favors and were made to do heavy labor while their men lived carefree lives. One tale described a woman having her ear bitten half off because she was slow completing the farm chores for her mate.

"John Adams and Ned Young, reversed the trend using the *Bounty* Bible as their textbook and guide for the settlement's return to the Christian religion and principled living. Young eventually died of asthma, leaving only Adams to continue the work of educating the women and children.

"Throughout the following years, several passing ships reported sighting human activity on the island, but in1808, Captain Mathew Folger and his crew on

---

[1.] 1. Adams (Alexander) was taught to read from the Bounty Bible by fellow mutineer, Ned Young. After Young's death, Adams built a school and taught the settlement's children to read and write along with following the Christian way of life. These spiritual roots later became the foundation for missionary efforts, which influenced the majority of the island's inhabitants to become Seventh Day Adventists by the end of the 19th Century. Today, that very Bible is housed under glass in the Pitcairn museum.

the *Topaz* came ashore and engaged in friendly conversation with Adams. As a result of a recommendation from the captain, Adams was granted amnesty by the Admiralty. The American sailing ship *Topaz* is recognized as the first to rediscover Pitcairn Island. After that, John Adams could freely have left the island but chose to remain in the place where he had spent half his life.

"His grave is the only one of a mutineer on the island that is marked with a headstone. As one might expect, the capital of the island is appropriately named Adamstown.

"The population of the island has fluctuated from lows of under 20 to as many as 233 since the mutineers first made it their home in the late 1700s. There have been on-going cycles of complete evacuation followed by periods of resettlement ever since. I was directly involved in one of the resettlements, the so-called Tahiti experiment. You will find my place in the sequence in the following chart of Important Pitcairn Events:

Important Pitcairn Events[1]

1. **July 1767** First European sighting of the island by Robert Pitcairn, a midshipman on *HMS Swallow*, and son of Major John Pitcairn of the Royal Marines, British commander at the first skirmishes of the American Revolutionary War. Captain Philip Carteret, *Swallow's* commander, calls it "Pitcairn's Island" in honor of Robert. (Young Pitcairn died at age 17 in 1769 when *HMS Aurora* was lost at sea.)

2. **December 1787** *HMAV Bounty*, under the command of William Bligh, sails from Spithead in England. Her mission is to collect breadfruit from Tahiti to be taken to the West Indies to feed slaves on plantations there.

3. **March 1788** Fletcher Christian, master's mate of the *Bounty*, is promoted to acting Second Lieutenant.

4. **October 26, 1788,** *HMAV Bounty* arrives at Tahiti. It will be at the island for 5 ½ months.

5. **April 28, 1789,** Under Fletcher Christian's leadership, many of the *Bounty's* sailors mutiny and cast Captain Bligh with 18 sailors loyal to him adrift in the ship's cutter.

6. **May 1789** The *Bounty*, with the mutineers sailing her, arrives at Tubuai but leaves after just three days at the island.

---

[1] *Important Pitcairn Events* adapted from Guide to Pitcairn with permission from The Pitcairn Council

7. **June 6, 1789,** Returning to Tahiti, the *Bounty* takes aboard a number of Polynesian men, women, and a baby. Livestock is also put aboard.

8. **June 23, 1789,** With its mutineer crew and Polynesians aboard, the Bounty returns to Tubuai.

9. **September 22, 1789,** The *Bounty* returns to Tahiti again. A decision is made to search for a safe hiding place. A total of 16 of the mutineers decide to take their chances at safety at Tahiti. Nine others, along with six Polynesian men, 12 women, and a baby girl, set sail to seek a safe haven.

10. **January 1790,** Pitcairn Island is sighted. After inspection of the island by Christian, it is decided to settle there. A factor in the decision is that the island has been misplaced on Admiralty maps and would thus be hard to find.

11. **January 23, 1790,** Whether by plan or by accident, *HMAV Bounty* is set afire. The unburned portion of the ship sinks in what is now called Bounty Bay.

12. **1790** Two of the Polynesian women brought to the island – Puarei and Tinafanaea – die. Two Polynesian men – Oha and Tararo – are murdered.

13. **September 1793,** In one day four of the mutineers – Fletcher Christian, John Mills, Isaac Martin, John Williams – are killed by Polynesian men.

14. **1798** Drunk on a locally distilled brew, mutineer William McCoy commits suicide.

15. **1799** Mutineers Edward Young and John Adams, believing their lives are in danger from the man, kill mutineer Matthew Quintal.

16. **December 1800** When Edward Young dies on Christmas Day, John Adams becomes the last of the mutineers alive on Pitcairn Island.

17. **February 6, 1808,** Pitcairn is re-discovered along with the mutineers' presence on the island by Captain Mayhew Folger of the American sealing ship *Topaz.*

18. **September 17, 1814,** *HMS Briton* and *HMS Tagus* unexpectedly call at Pitcairn. Captains of the ships correct the calendar error made when the *Bounty* crossed the international dateline.

19. **March 5, 1829,** John Adams, last of the mutineers on Pitcairn, dies. He is 65 years old. His wife, Teio (Mary) follows him in death nine days later.

20. **March 1831** The people of Pitcairn move to Tahiti for resettlement. Disease strikes quickly with death felling 12 of the Pitcairners, including Fletcher's eldest son, Thursday October Christian. The people decide to return to their Pitcairn home.

21. *__September 3, 1831,__ The 65 Pitcairners arrive at their island home on the American ship *Charles Doggett*, Captain William Driver. On his ship flies the U.S. "Stars and Stripes" flag to which Driver gives the name "*Old Glory*," a title that will be adopted nationally in America.

22. **June 1856** The Pitcairners – all 193 of them – immigrate to Norfolk Island – in the ship *Morayshire*. A baby is born during the voyage.

23. **1859** Homesick for Pitcairn, 16 of the Pitcairners are brought back to the island in the *Mary Ann*.

24. **1864** Four more families return to Pitcairn from Norfolk.[1]

---

[1] Although Captain Driver passed away at this point (item 24) in his journal, the authors chose to continue listing the significant Pitcairn events until the end of the nineteenth century.

25. 1890 The missionary schooner *Pitcairn*, with Adventist missionaries who would carry their faith to many islands of the Pacific aboard. Most of the adults on Pitcairn are baptized, formally making them members of the Adventist faith.

26. 1914 Pitcairn becomes a stopping point on the direct Panama to New Zealand shipping route when the Panama Canal opens.

27. 1940 The first issue of Pitcairn Islands stamps. Income from the worldwide sale of these popular postal adhesives will begin to fund operations and provide subsidies for Pitcairn and her people.

28. 1957 An anchor of *HMAV Bounty* is raised from Bounty Bay, Pitcairn Island.

29. January 1990 Celebrations marking the bicentennial of the settlement of Pitcairn are held.

30. January 8, 1999, What could be the last of the few cannons that were aboard *HMAV Bounty* is raised from the wreck site."

An opportunity to move to Pitcairn

Pitcairn Island is one of the most isolated places in the world, and in many respects, it is a tropical island paradise. According to their website, the government is committed to maintaining a vibrant community and attracting new migrants who wish to make a contribution to the community's sustainable future. Pitcairn's environment favours those who enjoy the outdoors, are at home in the natural, unspoilt environment, and would welcome the opportunity to be part of a small but lively community. Legislation now aims at ensuring that every permanent resident of Pitcairn Island shall be entitled to an allocation of house, garden, orchard and forestry land. Those interested in learning about this opportunity are invited to follow this link:

http://www.immigration.gov.pn/land/apply_for_land/index.html

I have been living alone since my Sarah passed away on September 1, 1878. It was seven years ago—the divine number of completion and perfection; that she was! My living children are scattered far and wide across the states, and some of them and I are estranged—no visits, talks, or family meals—not a word. That is an affliction I have not been able to reconcile. Perhaps I caused it, but deep down, when there is enmity between flesh and blood, the grief is most painful.

There are lots of letters I need to write to my surviving kin and especially to grandchildren I have not met and likely will never see. There is much to share about a most unusual one-of-a-kind life that will be lost forever if I don't write it down. I want them to know that concealed behind this facade of a difficult old man is boundless compassion, tenderness, and love for all people and my country. Please understand that my nature challenges me to show it...I must stop this reminiscing and get a rag to wipe my eyes; it must be pollen irritating them and causing a lump in my throat...

The old folks back through time have talked to me about death—the preparation, alerts, and warning signs. I too find my step slowing, and my feelings becoming more melancholy and thoughts more pensive. My legacy has crossed my mind, as well. No doubt everyone leaves one, regardless of how we wish it to be portrayed. I have plans soon to purchase my tombstone and prepare an inscription of parting words. What shall it say? There is much to consider.

I think back to my roots in Salem which have been transplanted to Nashville. I recall my forefathers in the Massachusetts Bay telling of hearing John Winthrop challenge them to figuratively create a magnificent City on a Hill that will be a positive example for others to follow—an inspiration. It seems more than a coincidence that the two greatest cities in my life—Salem, Massachusetts, and Nashville, Tennessee—have hills and bluffs all around. The people are so much alike—kindly, considerate, and down to earth. New folks who move in or come to visit never want to leave. As I consider how both have blessed me, but now I sense a premonition about the fleeting time I have remaining and an urgency to pen this message. Hopefully, it will serve to be my legacy—my symbolic City on a Hill—my modeling of patriotism.

To whom it may concern: *April 1, 1885*

This is a message to my grandchildren yet to come, the current citizens of the United States of America, and those who aspire to become citizens now and in the future.

My name is William Driver, an American Patriot, also known as Captain William Driver due to my many years commanding sailing ships to ports around the world. I first went to sea at the age of 14 and remained a mariner for 20 years.

At the age of 21, I was privileged and honored to have circumnavigated the entire world as Captain of the Charles Doggett. I do not make the claim, but historians have attributed to me—not Magellan— the distinction of being the youngest to have accomplished that feat at this juncture in history. Regardless of the correctness, I was able to achieve that odyssey twice.

As a merchant mariner, my voyages took me to the seven continents sailing on all the oceans and most of the seas; to the smallest cities in the world and to the largest; to lands that were unexplored and uncivilized as well as the most advanced of societies; and to remote islands harboring cannibals, headhunters, pirates, and savages. Consequently, I claim to be an eye witness authority on the cultures of almost every inhabited territory on earth. It is an easy choice for me to make a comparative analysis of the best places where I have lived and would recommend to others. I favor cities with hills where the boats can be seen off in the distance— blessed cities like Nashville and Salem and others from coast to coast and border to border.

Simply, I found The United States of America in its entirety to be that City on the Hill—the light of the world where the eyes of all people are upon it—Matthew 5:14. I see in my mind's eye displayed in its light a Constitution, a Bill of Rights and three branches of government that give us the assurance that all liberties will be guaranteed, all wrongs will be righted, and that justice always will prevail. There is something else—a certificate in the form of a glorious flag that waves as a warranty deed declaring all the foregoing to be true. I have called it "Old Glory" for 65 years, and the power and authority of its very presence have never failed me. It is the most comforting sight in the world for desperate eyes and the most feared symbol for the forces of evil and oppression. It has protected me in the most foreboding environments and circumstances.

This is how I show my appreciation for this country and recommend it as an honorable model for all others. Until death, I thankfully will praise God on my knees and our Country's flag standing on my feet with my hand respectfully over my heart. I have seen and copied a draft written by a Baptist minister's son that I am confident, in time, will become the prevailing expression of one's allegiance.

My legs are unsteady, and my hand is shaky. Will someone please help me to my feet one last time while I pay my respect? "I pledge allegiance to the Flag of the United States of America, and to the Republic for which it stands, one Nation indivisible, with liberty and justice for all."[1]

---

1. One nation [under God] was added in 1954 at the behest of President Eisenhower.

Old Glory [215]

William Driver, the definitive American patriot,
died March 2, 1886, in Nashville, Tennessee and is buried
in the Nashville City Cemetery. The site of the tomb
is one of the few in America that allows the display
of the American Flag 24 hours a day.

Epilogue

*T*his is Jack Benz again returning as myself after assuming the voice and playing the role of my great-great–grandfather during his 83-year odyssey. For emphasis, I recap the essence of my introductory remarks in my message to the readers.

Other accounts and books written by family members emerged, but his indirect role in the Academy Award best movie of the year in 1936—The Mutiny on the Bounty—made Captain William Driver a permanent fixture as a historical figure of great significance. As a result, newspaper and magazine articles became increasingly prolific as the Captain's fame and reputation widened and continued to grow.

As the years rolled by and new information accumulated, I found gaps, errors, omissions, and incompleteness in my lifelong frame of reference, which this effort has attempted to remedy and refine. There may be other errors, including those I unwittingly have contributed. In this, my first attempt as an author, I have taken the liberty to use a casual style without strict adherence to rigid formalities and protocols recommended for academic writing. However, careful attention has been applied to proper crediting every source utilized throughout. My most reliable resource is proprietary archived in my memory from the countless hours of word-of-mouth information transmitted from Grandmother Georgie.

Lastly, my co-author and I have attempted to bring a degree of our uniqueness to this writing. We hope you found it to be intriguing, informative, and entertaining; that said, there are two more significant events that occured after William Driver's passing. One is a defining moment in American History. The other is a defining moment in the lives of many thousands in the Driver bloodline.

Old Glory

"The apple doesn't fall far from the tree," generally is expressed and referenced to mean that children by heredity or environment tend to be like one or both of their parents. This seems to be the rule rather than the exception in terms of looks, intellect, character, and disposition. William Driver's immediate family had several members that fit in this

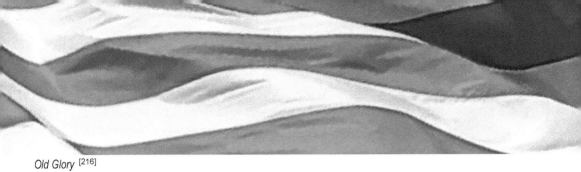

Old Glory [216]

mold quite accurately, especially in terms of their temperament.

Some tended to be quarrelsome, contentious, and stubborn—not at all unlike their father. There were factions and estrangements so intense that separation and isolation were perferred over engaging in conversation or acknowledging each other's existence. The potential for internal family rivalry was kindled due to there being children from different mothers living under one roof. Additionally, the family was torn apart because three of William's sons enlisted in the Confederacy, which in the father's thinking was akin to treason. In a letter in my possession dated March 1862, William wrote, "Two sons in the army of the South! My entire house estranged...and when I come home...no one to soothe me."

Foremost and of greatest consternation, his iconic flag provoked deep sibling and cousin conflict primarily between his daughter, Mary Jane Roland, and a niece, Harriet Ruth Waters Cooke. Ruth was the daughter of William's youngest sister, Harriet. In this narrative, there was mention of a special 12' X 24' American flag sewn from the finest merino and French lisse by William's mother and some neighborhood ladies. They presented it to him during a ceremony that honored his achieving the rank of Captain and becoming the commander of the *Charles Doggett*. It was a fitting 21st birthday gift as well. According to several reports, the ceremony followed a formal protocol for such an occasion, including a dedication and his hats off proclamation from William: "My Ship, my Country, and my Flag, *Old Glory*."

My Grandmother, Georgie Wade related that her mother Delilah "Dillie," recalled her Pa routinely quoting that phrase. Further evidence of his fondness for that specific expression is that those very words are included in the epitaph he wrote and had inscribed on his tombstone.

The Plot Thickens

At the end of the Civil War in 1865, William's eldest daughter, Mary Jane, married Union Officer Charles Roland and they moved out west. On a visit to her father and mother in 1873, William gave *Old Glory* to Mary Jane. He felt that she, being the oldest, was the logical one to inherit his pride and joy. She returned to Ogden, Utah after a four day and night train ride and a team of horses completed the trip to her home some 65 miles away.

That region of the country was experiencing a bonanza gold and platinum boom at that time. Two more of her adventuresome family, brothers Henry and Robert, were living there in a mining camp on the Salmon River in Nevada—a scenic paradise. She joined them on July 4, 1874, and displayed *Old Glory* for a joyous celebration including campers from every state in the Union. Mary Jane and Charles lived near the area in Wells, Nevada before retiring in Mountain View, California.

During that special July 4th celebration, Mary Jane shared with the gathering a striking comparison she heard directly from her father. It related to a different 4th of July that occurred during a cargo delivery en route from Australia to East India. As the ultimate patriot, he always celebrated Independence Day, whether it be in a foreign port, at sea, in Salem, or Nashville.

The destination was the Port of Calcutta, located about 125 miles up the Hooghly River from the Bay of Bengal on the Indian Ocean. The river is a branch of the Ganges, the holy river of the Hindu religion.

William told his daughter he never would forget the predicament he and his crew found themselves soon after their arrival. Although they encountered some complications related to international law, they were determined to carry out their traditional celebration. On the eve of the Fourth, the crew worked extra hard to unload most of the cargo. They finished in time to decorate the brig's rigging with every snip of red, white, and blue they could find onboard.

The next day all arose early to hoist *Old Glory* to the masthead, but their passion was stopped abruptly on orders from the harbormaster. Thanks to a stiff breeze, they were able to avoid further conflict by sailing beyond the three-mile limit where they completed

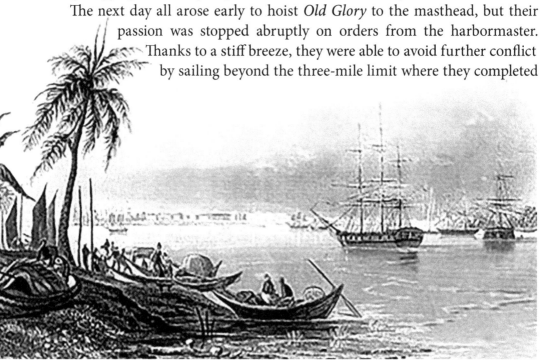

Calcutta [217]

their celebration without interference and anchored for the night.

The real blow came the next morning when they re-entered the harbor and learned that there was a horrible plague occurring on the Ganges. Stricken people were dying so fast that their bodies were floating like driftwood on the water. As soon as they moored at the port, they immediately were quarantined.

For sixty days, they were prevented from leaving, while fighting both starvation and the threat of disease. In desperation, Captain Driver decided to go ashore alone for provisions and information. He thoughtfully did not want any of his crew to be infected, so he learned about some crude methods of immunization which fortunately worked. When they finally were allowed to leave port, his ship was the only vessel that had not suffered a single death.

Like a maze, another *Old Glory* strangely emerged at the Essex Institute in Salem, Massachusetts. It had been donated in 1880 by William's niece, Harriet Ruth Waters Cooke, along with a note from William stating, "I send you this, my oldest flag…this is my original *Old Glory*, like me neglected and worn out." Obviously, this discovery enraged Mary Jane both at her cousin and the Essex Institute. She spent her remaining life disputing the legitimacy of the Essex flag and writing a book in 1908 supporting her rationale, "*Old Glory*: The True Story."

The Drivers were not at all averse to going to the head of the stream to right wrongs. William Driver personally knew and had conversed with three Presidents, i.e., Abe Lincoln, Andrew Johnson, and Rutherford B. Hayes, over various matters of concern. In like manner, Mary Jane contacted President Warren G. Harding and offered to donate the original *Old Glory*. He accepted and sent it to the Smithsonian Institute for posterity.

The authenticity of the two being claimed as the original was thoroughly reviewed and evaluated, producing a conclusion that the flag of Mary Jane Roland was the true *Old Glory*. The controversy ended with this verification, and the flag rests in peace at the Smithsonian in Washington, D.C. The numerous letters of documentation regarding the preceding are in my proprietary files collection at this time.

After Mary Jane was vindicated and had died, her husband, Charles, denounced Ruth Cooke's action as a cunning scheme. However, other peacemakers in the family gave her a conciliatory pass. My Great Grandmother, Delilah "Dillie" Reece noted that "Pa" (William) talked and wrote about many of his flags. Perhaps in his senility, he indeed sent Ruth one of them believing it was the original. Ruth had nothing to gain to misrepresent the flag since she donated it.

Nevertheless, I offer another perspective. It is not the materialistic item or its age that delivers the compelling message and redeeming value. It is the ideals symbolized by it and all Star-Spangled Banners—then and now.

A Surprising Revelation

Throughout this narrative, I was commissioned to speak in the voice of William Driver with full advance knowledge of what already had happened. I had the advantage of a retrospective view of his life and thought. I was able to speak on his behalf and reveal information for the readers that I did not have as myself. Therefore, it seems it is appropriate that I return as Jack Benz to help you understand the shock I experienced in September 1968.

My brother and I were visiting with our mother, Frances—a widow of 22 years—when she received a call from Louise Davis, a Nashville Tennessean writer of high distinction. We were familiar with her because she had written several interesting stories about my Great-Great-Grandfather, Captain William Driver. The reason for her call was quite shocking for two reasons. She had been interviewing a lawyer from New Zealand who claimed to be related to our family and who wanted to meet with us to discuss our relationship.

First of all, we had no knowledge of having a relative in New Zealand, and secondly, we were suspicious of his claim to be an attorney. My brother and I, both being in the insurance business, dealt with lawyers routinely and were somewhat uneasy because, in most instances involving lawyers, somebody was suing or being sued.

However, because Louise was motivated to contact us, we were sufficiently curious to agree to a meeting at the newspaper office the next day.

Rodney Acraman in 1968 [218]

My older brother, Billy, handled the arrangements and discussions on behalf of my mother and me. After a cordial introduction, we learned that the mystery guest's name was Rodney William Driver Acraman, a highly educated graduate of the University of New Zealand. He was an attorney representing the Fijian Government and was en route to London for official business and graduate study at Oxford University. He had traced the family ties through the Essex Institute in Salem, Massachusetts and was taking advantage of the travel circumstances to further his genealogical interests. No stated or implied connection to financial gain was apparent—strictly family history.

Our first impression heightened our curiosity because of his striking resemblance to numerous Captain Driver photographs we had collected through the years. He took the lead in presenting a very compelling account, beginning with his claim that we were his cousins due to the fact that we shared a common great-great-grandfather. Accordingly, this was initiated in mid-1820 due to a liaison between the William Driver and a Fijian beauty and daughter of an island king. Her name was Talei, and the union ultimately produced a son named Isaac Driver—Rodney's Great-Grandfather.

Rodney Acraman in 2006 [219]

In accord with island custom, it seemed that Talei was gifted to William in appreciation for a favor extended the king—perhaps a gun, watch, or something else highly scarce and cherished on the remote island. The king passed Talei's hand and a lock of her hair to William's hand, which he soon discovered to mean that they were married.

Acramen had no knowledge of the mutineer story. His information was that Captain Driver frequently visited as a trading merchant specializing in harvesting sea cucumbers as delineated in the earlier plot.

Isaac grew up to marry Elizabeth Simpson, a wealthy daughter of an English sea captain. He and his many descendants reputedly prospered in land ownership and money. According to Acraman, Isaac died about 1910 in his 90's but his progeny since propagated around the world into many thousands.

Although William Driver was not married at the time of this purported relationship and as one might expect, our saintly mother doubted the entire account, including the authenticity of Rodney Acraman. Good mothers always have a way of giving the most favorable spin on matters related to their children. However, Billy and I were convinced of his truthfulness. The circumstantial evidence, along with corroborating records, overwhelmingly supported his credibility and the accuracy of his testimony.

Unfortunately, there was no written language in Fiji at the time this narrative occurred. Therefore, we must depend upon word of mouth to reconstruct what censuses normally would reveal. However, due to marvels of the electronic age, our graphic's professional for this book, who also is a genealogical specialist, currently is in communication with Driver descendants from around the world—another story for another day. You are invited to visit our website – www.CaptainDriver.com – if you have any information regarding the Driver descendants or Driver ancestry line and to contribute further to this history and restoration.

Another Day

. . . AND THE STAR-SPANGLED BANNER
IN TRIUMPH DOTH WAVE, O'ER THE LAND OF THE
FREE AND THE HOME OF THE BRAVE; O THUS BE IT
EVER WHEN FREEMEN SHALL STAND . . . PRAISE
THE POWER THAT HATH MADE AND PRESERVED US
A NATION! THEN CONQUER WE MUST, WHEN OUR
CAUSE IT IS JUST, AND THIS BE OUR MOTTO - "IN GOD
IS OUR TRUST," AND THE STAR-SPANGLED BANNER IN
TRIUMPH SHALL WAVE O'ER THE LAND OF THE FREE
AND THE HOME OF THE BRAVE.

Old Glory [220]

Post Script

I close this odyssey with special recognition, appreciation, and acknowledgment of the talented two who helped me tell this story. They specifically were highlighted on page 273.

For years, I took the path of least resistance on this project, which produced nothing—good intentions indeed but overwhelmed by tasks not understood and inadequate know-how. I had the ingredients in terms of materials, resources, and willingness but lacked the skills and confidence and guidance to travel a path. Thankfully, they patiently encouraged me to continue.

My motivation came from a feeling of responsibility on behalf of my great-great-grandfather to complete his unfinished business. There were missing links, unpublished documents, corrections, and unknown facts about his story—refinements that are extremely important to his legacy; and to our nation crying for his example of unwavering patriotism and for a rebuff of those who show it disrespect. I proudly salute you and your message, Captain William Driver!

I offer a special thanks to Noroma and Bentley for their extreme patience and support. This undertaking has been a distraction from our family's customary routine—a burden they have had to endure; and finally to countless friends whose daily encouragement and nudging gave me the desire to continue what I wanted to do but never before found the resolve—this was my last chance "to sail on."

Quod Erat Demonstrandum

Appendix

Included in this Appendix is a variety of reference resources which are captioned and are self-explanatory. The founding father documents among them merit further remarks. Readers born during the first half of the 20th century readily will recall those times of challenge and the tremendous patriotic response. The younger generations likely will find this flashback to be a startling revelation.

Buy Bonds Poster [221]

Most families were struggling as our country was in the ten-year recovery stage following the Great Depression of 1929. The Second World War was imminent or already had commenced. A record number of women went to work at various plants to support the war effort. Rationing of food, clothing, and gasoline was underway along with widespread collecting of scrap metal and other materials to support the war effort. Every boy and girl regularly saved their pennies to buy government "Savings Stamps" with pictures of a minuteman—ready to defend in a minute—at the cost of ten cents each. Ultimately, they could be traded for "Savings Bonds" and redeemed after ten years for $25 for each $18.75 invested. Every man, woman, and child was on board fully committed with love, money, and service to our country. It was unquestioned patriotism.

Scene at the Signing of the Constitution of the United States [122]

Our school days began with us standing by our desks before the Star-Spangled Banner with hands on our hearts and reciting the Pledge of Allegiance. Afterward, there was a nonsectarian Bible reading, the singing of the National Anthem. This routine was a pattern that continued for decades and generations.

Outward displays of patriotism tend to peak in times of trouble and challenge and wane somewhat when things are going well. This was evidenced at the time of the American Revolution and the series of wars that followed. The reaction after September 11, 2001, to the Twin Towers attack on our home soil, generated the highest display of patriotic togetherness in modern times.

Now, less than two decades later, a 2018 Gallup Poll survey revealed that only 47 percent of those surveyed are extremely proud to be Americans. I urge readers to access on-line links to the endless political correctness movements thwarting the promotion of patriotic practices particularly in the education institutions of our nation. The State of the Union appears to be in serious need of being revitalized in terms of understanding the ideals and values upon which our country was founded and has thrived. Our very survival is at stake. Hopefully, the foundation documents herein, along with the examples of Captain William Driver's principled life will motivate each reader to promote patriotism at every opportunity.

The Four Charters of Freedom

Foundation principles and philosophy of the United States of America.

1. Declaration of Independence (1776).
2. Constitution of the United States (1787) and subsequent Amendments
3. Bill of Rights (1791)
4. Federalist/Anti-Federalist Papers.

The Principal Founding Fathers

George Washington	Alexander Hamilton
Benjamin Franklin	John Adams
Samuel Adams	Thomas Jefferson
James Madison	John Jay

Historic plaque at Grace Church Newark [223]

Other Important Founders

Many other figures have also been cited as Founding Fathers (or Mothers). These include John Hancock, best known for his flashy signature on the Declaration of Independence; Gouverneur Morris, who wrote much of the Constitution; Thomas Paine, the British-born author of Common Sense; Paul Revere, a Boston silversmith whose "midnight ride" warned of approaching redcoats; George Mason, who helped craft the Constitution but ultimately refused to sign it; Charles Carroll, the lone Catholic to sign the Declaration of Independence; Patrick Henry, who famously declared, "Give me liberty, or give me death!"; John Marshall, a Revolutionary War veteran and longtime chief justice of the Supreme Court; and Abigail Adams, who implored her husband, John, to "remember the ladies" while shaping the new country.

Online links to the full text of the major foundation documents:

1. https://founders.archives.gov/
2. www.u.s.history.org
3. https://www.constitutionfacts.com/us-founding-fathers/
4. https://nccs.net/blogs/americas-founding-documents/americas-founding-documents
5. https://www.congress.gov/resources/display/content/Resources+A+to+Z

[1] Follow this link for text: https://www.congress.gov/resources/display/content/The+Federalist+

[224]

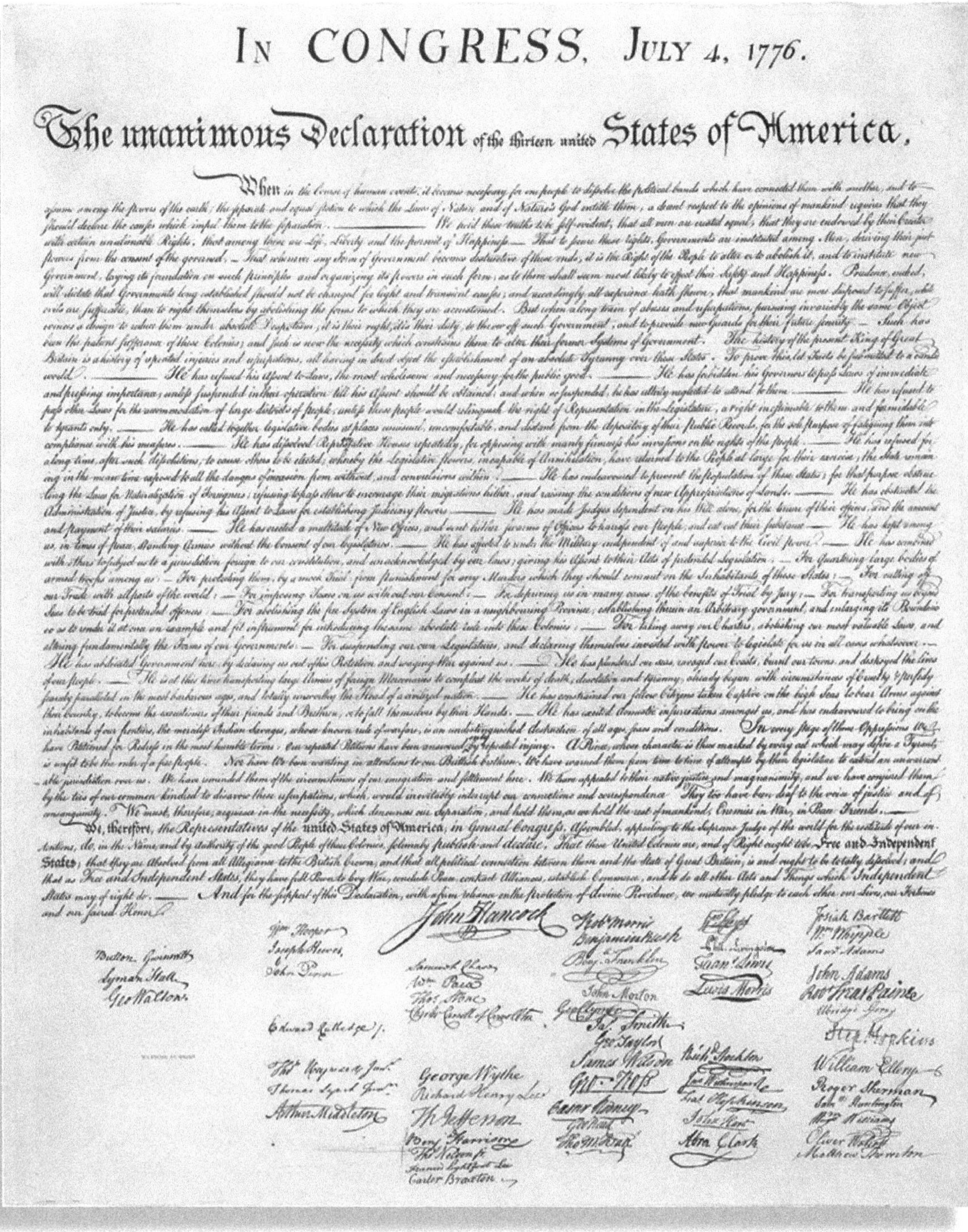

In Congress, July 4, 1776.

The unanimous Declaration of the thirteen united States of America,

When in the Course of human events, it becomes necessary for one people to dissolve the political bands which have connected them with another, and to assume among the powers of the earth, the separate and equal station to which the Laws of Nature and of Nature's God entitle them, a decent respect to the opinions of mankind requires that they should declare the causes which impel them to the separation.

We hold these truths to be self-evident, that all men are created equal, that they are endowed by their Creator with certain unalienable Rights, that among these are Life, Liberty and the pursuit of Happiness.

-- That to secure these rights, Governments are instituted among Men, deriving their just powers from the consent of the governed, --That whenever any Form of Government becomes destructive of these ends, it is the Right of the People to alter or to abolish it, and to institute new Government, laying its foundation on such principles and organizing its powers in such form, as to them shall seem most likely to effect their Safety and Happiness. Prudence, indeed, will dictate that Governments long established should not be changed for light and transient causes; and accordingly all experience hath shewn, that mankind are more disposed to suffer, while evils are sufferable, than to right themselves by abolishing the forms to which they are accustomed. But when a long train of abuses and usurpations, pursuing invariably the same Object evinces a design to reduce them under absolute Despotism, it is their right, it is their duty, to throw off such Government, and to provide new Guards for their future security.--Such has been the patient sufferance of these Colonies; and such is now the necessity which constrains them to alter their former Systems of Government. The history of the present King of Great Britain is a history of repeated injuries and usurpations, all having in direct object the establishment of an absolute Tyranny over these States. To prove this, let Facts be submitted to a candid world.

He has refused his Assent to Laws, the most wholesome and necessary for the public good.

He has forbidden his Governors to pass Laws of immediate and pressing importance, unless suspended in their operation till his Assent should be obtained; and when so suspended, he has utterly neglected to attend to them.

He has refused to pass other Laws for the accommodation of large districts of people, unless those people would relinquish the right of Representation in the Legislature, a right inestimable to them and formidable to tyrants only.

He has called together legislative bodies at places unusual, uncomfortable, and distant from the depository of their public Records, for the sole purpose of fatiguing them into compliance with his measures.

He has dissolved Representative Houses repeatedly, for opposing with manly firmness his invasions on the rights of the people.

He has refused for a long time, after such dissolutions, to cause others to be elected; whereby the Legislative powers, incapable of Annihilation, have returned to the People at large for their exercise; the State remaining in the mean time exposed to all the dangers of invasion from without, and convulsions within.

He has endeavoured to prevent the population of these States; for that purpose obstructing the Laws for Naturalization of Foreigners; refusing to pass others to encourage their migrations hither, and raising the conditions of new Appropriations of Lands.

He has obstructed the Administration of Justice, by refusing his Assent to Laws for establishing Judiciary powers.

He has made Judges dependent on his Will alone, for the tenure of their offices, and the amount and payment of their salaries.

He has erected a multitude of New Offices, and sent hither swarms of Officers to harrass our people, and eat out their substance.

He has kept among us, in times of peace, Standing Armies without the Consent of our legislatures.

He has affected to render the Military independent of and superior to the Civil power.

He has combined with others to subject us to a jurisdiction foreign to our constitution, and unacknowledged by our laws; giving his Assent to their Acts of pretended Legislation:

For Quartering large bodies of armed troops among us:

For protecting them, by a mock Trial, from punishment for any Murders which they should commit on the Inhabitants of these States:

For cutting off our Trade with all parts of the world:

For imposing Taxes on us without our Consent:

For depriving us in many cases, of the benefits of Trial by Jury:

For transporting us beyond Seas to be tried for pretended offences

For abolishing the free System of English Laws in a neighbouring Province, establishing therein an Arbitrary government, and enlarging its Boundaries so as to render it at once an example and fit instrument for introducing the same absolute rule into these Colonies:

For taking away our Charters, abolishing our most valuable Laws, and altering fundamentally the Forms of our Governments:

For suspending our own Legislatures, and declaring themselves invested with power to legislate for us in all cases whatsoever.

He has abdicated Government here, by declaring us out of his Protection and waging War against us.

He has plundered our seas, ravaged our Coasts, burnt our towns, and destroyed the lives of our people.

He is at this time transporting large Armies of foreign Mercenaries to compleat the works of death, desolation and tyranny, already begun with circumstances of Cruelty & perfidy scarcely paralleled in the most barbarous ages, and totally unworthy the Head of a civilized nation.

He has constrained our fellow Citizens taken Captive on the high Seas to bear Arms against their Country, to become the executioners of their friends and Brethren, or to fall themselves by their Hands.

He has excited domestic insurrections amongst us, and has endeavoured to bring on the inhabitants of our frontiers, the merciless Indian Savages, whose known rule of warfare, is an undistinguished destruction of all ages, sexes and conditions.

In every stage of these Oppressions We have Petitioned for Redress in the most humble terms: Our repeated Petitions have been answered only by repeated injury. A Prince whose character is thus marked by every act which may define a Tyrant, is unfit to be the ruler of a free people.

Nor have We been wanting in attentions to our Brittish brethren. We have warned them from time to time of attempts by their legislature to extend an unwarrantable jurisdiction over us. We have reminded them of the circumstances of our emigration and settlement here. We have appealed to their native justice and magnanimity, and we have conjured them by the ties of our common kindred to disavow these usurpations, which, would inevitably interrupt our connections and correspondence. They too have been deaf to the voice of justice and of consanguinity. We must, therefore, acquiesce in the necessity, which denounces our Separation, and hold them, as we hold the rest of mankind, Enemies in War, in Peace Friends.

We, therefore, the Representatives of the united States of America, in General Congress, Assembled, appealing to the Supreme Judge of the world for the rectitude of our intentions, do, in the Name, and by Authority of the good People of these Colonies, solemnly publish and declare, That these United Colonies are, and of Right ought to be Free and Independent States; that they are Absolved from all Allegiance to the British Crown, and that all political connection between them and the State of Great Britain, is and ought to be totally dissolved; and that as Free and Independent States, they have full Power to levy War, conclude Peace, contract Alliances, establish Commerce, and to do all other Acts and Things which Independent States may of right do. And for the support of this Declaration, with a firm reliance on the protection of divine Providence, we mutually pledge to each other our Lives, our Fortunes and our sacred Honor.

The 56 signatures on the Declaration appear in the positions indicated:

GEORGIA:
Button Gwinnett, Lyman Hall, George Walton
NORTH CAROLINA:
William Hooper, Joseph Hewes, John Penn
SOUTH CAROLINA:
Edward Rutledge, Thomas Heyward, Jr., Thomas Lynch, Jr., Arthur Middleton
MASSACHUSETTS:
John Hancock
MARYLAND:
Samuel Chase, William Paca, Thomas Stone, Charles Carroll of Carrollton
VIRGINIA:
George Wythe, Richard Henry Lee, Thomas Jefferson, Benjamin Harrison, Thomas Nelson, Jr., Francis Lightfoot Lee, Carter Braxton
PENNSYLVANIA:
Robert Morris, Benjamin Rush, Benjamin Franklin, John Morton, George Clymer, James Smith, George Taylor, James Wilson, George Ross
DELAWARE:
Caesar Rodney, George Read, Thomas McKean
NEW YORK:
William Floyd, Philip Livingston, Francis Lewis, Lewis Morris
NEW JERSEY:
Richard Stockton, John Witherspoon, Francis Hopkinson, John Hart, Abraham Clark
NEW HAMPSHIRE:
Josiah Bartlett, William Whipple
MASSACHUSETTS:
Samuel Adams, John Adams, Robert Treat Paine, Elbridge Gerry
RHODE ISLAND:
Stephen Hopkins, William Ellery
CONNECTICUT:
Roger Sherman, Samuel Huntington, William Williams, Oliver Wolcott
NEW HAMPSHIRE:
Matthew Thornton

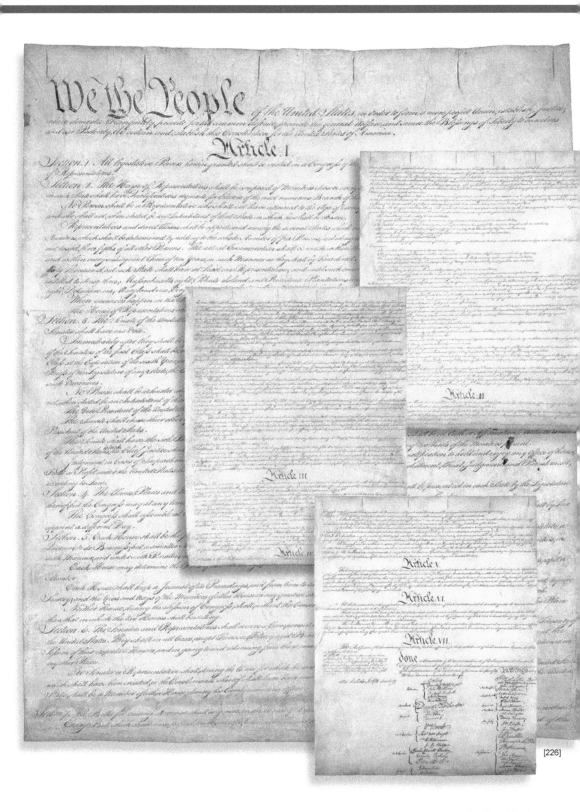

[226]

We the people of the United States, in order to form a more perfect union, establish justice, insure domestic tranquility, provide for the common defense, promote the general welfare, and secure the blessings of liberty to ourselves and our posterity, do ordain and establish this Constitution for the United States of America.

Article I

SECTION 1. All legislative powers herein granted shall be vested in a Congress of the United States, which shall consist of a Senate and House of Representatives.

SECTION 2. The House of Representatives shall be composed of members chosen every second year by the people of the several states, and the electors in each state shall have the qualifications requisite for electors of the most numerous branch of the state legislature.

No person shall be a Representative who shall not have attained to the age of twenty five years, and been seven years a citizen of the United States, and who shall not, when elected, be an inhabitant of that state in which he shall be chosen.

Representatives and direct taxes shall be apportioned among the several states which may be included within this union, according to their respective numbers, which shall be determined by adding to the whole number of free persons, including those bound to service for a term of years, and excluding Indians not taxed, three fifths of all other Persons. The actual Enumeration shall be made within three years after the first meeting of the Congress of the United States, and within every subsequent term of ten years, in such manner as they shall by law direct. The number of Representatives shall not exceed one for every thirty thousand, but each state shall have at least one Representative; and until such enumeration shall be made, the state of New Hampshire shall be entitled to chuse three, Massachusetts eight, Rhode Island and Providence Plantations one, Connecticut five, New York six, New Jersey four, Pennsylvania eight, Delaware one, Maryland six, Virginia ten, North Carolina five, South Carolina five, and Georgia three.

When vacancies happen in the Representation from any state, the executive authority thereof shall issue writs of election to fill such vacancies.

The House of Representatives shall choose their speaker and other officers; and shall have the sole power of impeachment.

SECTION 3. The Senate of the United States shall be composed of two Senators from each state, chosen by the legislature thereof, for six years; and each Senator shall have one vote.

Immediately after they shall be assembled in consequence of the first election, they shall be divided as equally as may be into three classes. The seats of the Senators of the first class shall be vacated at the expiration of the second year, of the second class at the expiration of the fourth year, and the third class at the expiration of the sixth year, so that one third may be chosen every second year; and if vacancies happen by resignation, or otherwise, during the recess of the legislature of any state, the executive thereof may make temporary appointments until the next meeting of the legislature, which shall then fill such vacancies.

No person shall be a Senator who shall not have attained to the age of thirty years, and been nine years a citizen of the United States and who shall not, when elected, be an inhabitant of that state for which he shall be chosen.

The Vice President of the United States shall be President of the Senate, but shall have no vote, unless they be equally divided.

The Senate shall choose their other officers, and also a President pro tempore, in the absence of the Vice President, or when he shall exercise the office of President of the United States.

The Senate shall have the sole power to try all impeachments. When sitting for that purpose, they shall be on oath or affirmation. When the President of the United States is tried, the Chief Justice shall preside: And no person shall be convicted without the concurrence of two thirds of the members present.

Judgment in cases of impeachment shall not extend further than to removal from office, and disqualification to hold and enjoy any office of honor, trust or profit under the United States: but the party convicted shall nevertheless be liable and subject to indictment, trial, judgment and punishment, according to law.

SECTION 4. The times, places and manner of holding elections for Senators and Representatives, shall be prescribed in each state by the legislature thereof; but the Congress may at any time by law make or alter such regulations, except as to the places of choosing Senators. The Congress shall assemble at least once in every year, and such meeting shall be on the first Monday in December, unless they shall by law appoint a different day.

SECTION 5. Each House shall be the judge of the elections, returns and qualifications of its own members, and a majority of each shall constitute a quorum to do business; but a smaller number may adjourn from day to day, and may be authorized to compel the attendance of absent members, in such manner, and under such penalties as each House may provide.

Each House may determine the rules of its proceedings, punish its members for disorderly behavior, and, with the concurrence of two thirds, expel a member.

Each House shall keep a journal of its proceedings, and from time to time publish the same, excepting such parts as may in their judgment require secrecy; and the yeas and nays of the members of either House on any question shall, at the desire of one fifth of those present, be entered on the journal.

Neither House, during the session of Congress, shall, without the consent of the other, adjourn for more than three days, nor to any other place than that in which the two Houses shall be sitting.

SECTION 6. The Senators and Representatives shall receive a compensation for their services, to be ascertained by law, and paid out of the treasury of the United States. They shall in all cases, except treason, felony and breach of the peace, be privileged from arrest during their attendance at the session of their respective Houses, and in going to and returning from the same; and for any speech or debate in either House, they shall not be questioned in any other place.

No Senator or Representative shall, during the time for which he was elected, be appointed to any civil office under the authority of the United States, which shall have been created, or the emoluments whereof shall have been increased during such time: and no person holding any office under the United States, shall be a member of either House during his continuance in office.

SECTION 7. All bills for raising revenue shall originate in the House of Representatives; but the Senate may propose or concur with amendments as on other Bills.

Every bill which shall have passed the House of Representatives and the Senate, shall, before it become a law, be presented to the President of the United States; if he approve he shall sign it, but if not he shall return it, with his objections to that House in which it shall have originated, who shall enter the objections at large on their journal, and proceed to reconsider it. If after such reconsideration two thirds of that House shall agree to pass the bill, it shall be sent, together with the objections, to the other House, by which it shall likewise be reconsidered, and if approved by two thirds of that House, it shall become a law. But in all such cases the votes of both Houses shall be determined by yeas and nays, and the names of the persons voting for and against the bill shall be entered on the journal of each House respectively. If any bill shall not be returned by the President within ten days (Sundays excepted) after it shall have been presented to him, the same shall be a law, in like manner as if he had signed it, unless the Congress by their adjournment prevent its return, in which case it shall not be a law.

Every order, resolution, or vote to which the concurrence of the Senate and House of

Representatives may be necessary (except on a question of adjournment) shall be presented to the President of the United States; and before the same shall take effect, shall be approved by him, or being disapproved by him, shall be repassed by two thirds of the Senate and House of Representatives, according to the rules and limitations prescribed in the case of a bill.

SECTION 8. The Congress shall have power to lay and collect taxes, duties, imposts and excises, to pay the debts and provide for the common defense and general welfare of the United States; but all duties, imposts and excises shall be uniform throughout the United States;

To borrow money on the credit of the United States;

To regulate commerce with foreign nations, and among the several states, and with the Indian tribes;

To establish a uniform rule of naturalization, and uniform laws on the subject of bankruptcies throughout the United States;

To coin money, regulate the value thereof, and of foreign coin, and fix the standard of weights and measures;

To provide for the punishment of counterfeiting the securities and current coin of the United States;

To establish post offices and post roads;

To promote the progress of science and useful arts, by securing for limited times to authors and inventors the exclusive right to their respective writings and discoveries;

To constitute tribunals inferior to the Supreme Court;

To define and punish piracies and felonies committed on the high seas, and offenses against the law of nations;

To declare war, grant letters of marque and reprisal, and make rules concerning captures on land and water;

To raise and support armies, but no appropriation of money to that use shall be for a longer term than two years;

To provide and maintain a navy;

To make rules for the government and regulation of the land and naval forces;

To provide for calling forth the militia to execute the laws of the union, suppress insurrections and repel invasions;

To provide for organizing, arming, and disciplining, the militia, and for governing such part of them as may be employed in the service of the United States, reserving to the states respectively, the appointment of the officers, and the authority of training the militia according to the discipline prescribed by Congress;

To exercise exclusive legislation in all cases whatsoever, over such District (not exceeding ten miles square) as may, by cession of particular states, and the acceptance of Congress, become the seat of the government of the United States, and to exercise like authority over all places purchased by the consent of the legislature of the state in which the same shall be, for the erection of forts, magazines, arsenals, dockyards, and other needful buildings;-- And

To make all laws which shall be necessary and proper for carrying into execution the foregoing powers, and all other powers vested by this Constitution in the government of the United States, or in any department or officer thereof.

SECTION 9. The migration or importation of such persons as any of the states now existing shall think proper to admit, shall not be prohibited by the Congress prior to the year one thousand eight hundred and eight, but a tax or duty may be imposed on such importation, not exceeding ten dollars for each person.

The privilege of the writ of habeas corpus shall not be suspended, unless when in cases of rebellion or invasion the public safety may require it.

No bill of attainder or ex post facto Law shall be passed.

No capitation, or other direct, tax shall be laid, unless in proportion to the census or enumeration herein before directed to be taken.

No tax or duty shall be laid on articles exported from any state.

No preference shall be given by any regulation of commerce or revenue to the ports of one state over those of another: nor shall vessels bound to, or from, one state, be obliged to enter, clear or pay duties in another.

No money shall be drawn from the treasury, but in consequence of appropriations made by law; and a regular statement and account of receipts and expenditures of all public money shall be published from time to time.

No title of nobility shall be granted by the United States: and no person holding any office of profit or trust under them, shall, without the consent of the Congress, accept of any present, emolument, office, or title, of any kind whatever, from any king, prince, or foreign state.

SECTION 10. No state shall enter into any treaty, alliance, or confederation; grant letters of marque and reprisal; coin money; emit bills of credit; make anything but gold and silver coin a tender in payment of debts; pass any bill of attainder, ex post facto law, or law impairing the obligation of contracts, or grant any title of nobility.

No state shall, without the consent of the Congress, lay any imposts or duties on imports or exports, except what may be absolutely necessary for executing its inspection laws: and the net produce of all duties and imposts, laid by any state on imports or exports, shall be for the use of the treasury of the United States; and all such laws shall be subject to the revision and control of the Congress.

No state shall, without the consent of Congress, lay any duty of tonnage, keep troops, or ships of war in time of peace, enter into any agreement or compact with another state, or with a foreign power, or engage in war, unless actually invaded, or in such imminent danger as will not admit of delay.

Article II

SECTION 1. The executive power shall be vested in a President of the United States of America. He shall hold his office during the term of four years, and, together with the Vice President, chosen for the same term, be elected, as follows:

Each state shall appoint, in such manner as the Legislature thereof may direct, a number of electors, equal to the whole number of Senators and Representatives to which the State may be entitled in the Congress: but no Senator or Representative, or person holding an office of trust or profit under the United States, shall be appointed an elector.

The electors shall meet in their respective states, and vote by ballot for two persons, of whom one at least shall not be an inhabitant of the same state with themselves. And they shall make a list of all the persons voted for, and of the number of votes for each; which list they shall sign and certify, and transmit sealed to the seat of the government of the United States, directed to the President of the Senate. The President of the Senate shall, in the presence of the Senate and House of Representatives, open all the certificates, and the votes shall then be counted. The person having the greatest number of votes shall be the President, if such number be a majority of the whole number of electors appointed; and if there be more than one who have such majority, and have an equal number of votes, then the House of Representatives shall immediately choose by ballot one of them for President; and if no person have a majority, then from the five highest on the list the said House shall in like manner choose the President. But in choosing the President, the votes shall be taken by States, the representation from each state having one vote; A quorum for this purpose shall consist of a member or members from two thirds of the states, and a majority of all the states shall be necessary to a choice. In every case, after the choice of the President, the person having the greatest number of votes of the electors shall be the Vice President. But if there should remain two or more who have equal votes, the Senate shall choose from them by ballot the Vice President.

The Congress may determine the time of choosing the electors, and the day on which they shall give their votes; which day shall be the same throughout the United States.

No person except a natural born citizen, or a citizen of the United States, at the time of the adoption of this Constitution, shall be eligible to the office of President; neither shall any person be eligible to that office who shall not have attained to the age of thirty five years, and been fourteen Years a resident within the United States.

In case of the removal of the President from office, or of his death, resignation, or inability to discharge the powers and duties of the said office, the same shall devolve on the Vice President, and the Congress may by law provide for the case of removal, death, resignation or inability, both of the President and Vice President, declaring what officer shall then act as President, and such officer shall act accordingly, until the disability be removed, or a President shall be elected.

The President shall, at stated times, receive for his services, a compensation, which shall neither be increased nor diminished during the period for which he shall have been elected, and he shall not receive within that period any other emolument from the United States, or any of them.

Before he enter on the execution of his office, he shall take the following oath or affirmation:-- "I do solemnly swear (or affirm) that I will faithfully execute the office of President of the United States, and will to the best of my ability, preserve, protect and defend the Constitution of the United States."

SECTION 2. The President shall be commander in chief of the Army and Navy of the United States, and of the militia of the several states, when called into the actual service of the United States; he may require the opinion, in writing, of the principal officer in each of the executive departments, upon any subject relating to the duties of their respective offices, and he shall have power to grant reprieves and pardons for offenses against the United States, except in cases of impeachment.

He shall have power, by and with the advice and consent of the Senate, to make treaties, provided two thirds of the Senators present concur; and he shall nominate, and by and with the advice and consent of the Senate, shall appoint ambassadors, other public ministers and consuls, judges of the Supreme Court, and all other officers of the United States, whose appointments are not herein otherwise provided for, and which shall be established by law: but the Congress may by law vest the appointment of such inferior officers, as they think proper, in the President alone, in the courts of law, or in the heads of departments.

The President shall have power to fill up all vacancies that may happen during the recess of the Senate, by granting commissions which shall expire at the end of their next session.

SECTION 3. He shall from time to time give to the Congress information of the state of the union, and recommend to their consideration such measures as he shall judge necessary and expedient; he may, on extraordinary occasions, convene both Houses, or either of them, and in case of disagreement between them, with respect to the time of adjournment, he may adjourn them to such time as he shall think proper; he shall receive ambassadors and other public ministers; he shall take care that the laws be faithfully executed, and shall commission all the officers of the United States.

SECTION 4. The President, Vice President and all civil officers of the United States, shall be removed from office on impeachment for, and conviction of, treason, bribery, or other high crimes and misdemeanors.

Article III

SECTION 1. The judicial power of the United States, shall be vested in one Supreme Court, and in such inferior courts as the Congress may from time to time ordain and establish. The judges, both of the supreme and inferior courts, shall hold their offices during good behaviour, and shall, at stated times, receive for their services, a compensation, which shall not be diminished during their continuance in office.

SECTION 2. The judicial power shall extend to all cases, in law and equity, arising under this

Constitution, the laws of the United States, and treaties made, or which shall be made, under their authority;- -to all cases affecting ambassadors, other public ministers and consuls;--to all cases of admiralty and maritime jurisdiction;--to controversies to which the United States shall be a party;--to controversies between two or more states;--between a state and citizens of another state;-- between citizens of different states;--between citizens of the same state claiming lands under grants of different states, and between a state, or the citizens thereof, and foreign states, citizens or subjects.

In all cases affecting ambassadors, other public ministers and consuls, and those in which a state shall be party, the Supreme Court shall have original jurisdiction. In all the other cases before mentioned, the Supreme Court shall have appellate jurisdiction, both as to law and fact, with such exceptions, and under such regulations as the Congress shall make.

The trial of all crimes, except in cases of impeachment, shall be by jury; and such trial shall be held in the state where the said crimes shall have been committed; but when not committed within any state, the trial shall be at such place or places as the Congress may by law have directed.

SECTION 3. Treason against the United States, shall consist only in levying war against them, or in adhering to their enemies, giving them aid and comfort. No person shall be convicted of treason unless on the testimony of two witnesses to the same overt act, or on confession in open court.

The Congress shall have power to declare the punishment of treason, but no attainder of treason shall work corruption of blood, or forfeiture except during the life of the person attainted.

Article IV

SECTION 1. Full faith and credit shall be given in each state to the public acts, records, and judicial proceedings of every other state. And the Congress may by general laws prescribe the manner in which such acts, records, and proceedings shall be proved, and the effect thereof.

SECTION 2. The citizens of each state shall be entitled to all privileges and immunities of citizens in the several states.

A person charged in any state with treason, felony, or other crime, who shall flee from justice, and be found in another state, shall on demand of the executive authority of the state from which he fled, be delivered up, to be removed to the state having jurisdiction of the crime.

No person held to service or labor in one state, under the laws thereof, escaping into another, shall, in consequence of any law or regulation therein, be discharged from such service or labor, but shall be delivered up on claim of the party to whom such service or labor may be due.

SECTION 3. New states may be admitted by the Congress into this union; but no new states shall be formed or erected within the jurisdiction of any other state; nor any state be formed by the junction of two or more states, or parts of states, without the consent of the legislatures of the states concerned as well as of the Congress.

The Congress shall have power to dispose of and make all needful rules and regulations respecting the territory or other property belonging to the United States; and nothing in this Constitution shall be so construed as to prejudice any claims of the United States, or of any particular state.

SECTION 4. The United States shall guarantee to every state in this union a republican form of government, and shall protect each of them against invasion; and on application of the legislature, or of the executive (when the legislature cannot be convened) against domestic violence.

Article V

The Congress, whenever two thirds of both houses shall deem it necessary, shall propose amendments to this Constitution, or, on the application of the legislatures of two thirds of the several states, shall call a convention for proposing amendments, which, in either case, shall be valid to all intents and purposes, as part of this Constitution, when ratified by the legislatures of three fourths of the several states, or by conventions in three fourths thereof, as the one or the other mode of ratification may be proposed by the Congress; provided that no amendment which may be made prior to the year one thousand eight hundred and eight shall in any manner affect the first and fourth clauses in the ninth section of the first article; and that no state, without its consent, shall be deprived of its equal suffrage in the Senate.

Article VI

All debts contracted and engagements entered into, before the adoption of this Constitution, shall be as valid against the United States under this Constitution, as under the Confederation.

This Constitution, and the laws of the United States which shall be made in pursuance thereof; and all treaties made, or which shall be made, under the authority of the United States, shall be the supreme law of the land; and the judges in every state shall be bound thereby, anything in the Constitution or laws of any State to the contrary notwithstanding.

The Senators and Representatives before mentioned, and the members of the several state legislatures, and all executive and judicial officers, both of the United States and of the several states, shall be bound by oath or affirmation, to support this Constitution; but no religious test shall ever be required as a qualification to any office or public trust under the United States.

Article VII

The ratification of the conventions of nine states, shall be sufficient for the establishment of this Constitution between the states so ratifying the same.

Done in convention by the unanimous consent of the states present the seventeenth day of September in the year of our Lord one thousand seven hundred and eighty seven and of the independence of the United States of America the twelfth. In witness whereof We have hereunto subscribed our Names,

G. Washington-Presidt. and deputy from Virginia

NEW HAMPSHIRE

John Langdon Nicholas Gilman

MASSACHUSETTS

Nathaniel Gorham Rufus King

CONNECTICUT

Wm: Saml. Johnson Roger Sherman

NEW YORK

Alexander Hamilton

NEW JERSEY

Wil: Livingston David Brearly Wm. Paterson Jona: Dayton

PENNSYLVANIA

B. Franklin Thomas Mifflin Robt. Morris Geo. Clymer Thos. FitzSimons Jared Ingersoll James Wilson Gouv Morris

DELAWARE

Geo: Read Gunning Bedford jun John Dickinson Richard Bassett Jaco: Broom

MARYLAND

James McHenry Dan of St Thos. Jenifer Danl Carroll

VIRGINIA

John Blair— James Madison Jr.

NORTH CAROLINA

Wm. Blount Richd. Dobbs Spaight Hu Williamson

SOUTH CAROLINA

J. Rutledge Charles Cotesworth Pinckney Charles Pinckney Pierce Butler

GEORGIA

William Few Abr Baldwin

Bill of Rights

Note: The following text is a transcription of the first ten amendments to the Constitution in their original form. These amendments were ratified December 15, 1791, and form what is known as the "Bill of Rights."

[227]

The Preamble to The Bill of Rights
Congress of the United States
begun and held at the City of New York, on-
Wednesday the fourth of March, one thousand seven hundred and eighty nine.

THE Conventions of a number of the States, having at the time of their adopting the Constitution, expressed a desire, in order to prevent misconstruction or abuse of its powers, that further declaratory and restrictive clauses should be added: And as extending the ground of public confidence in the Government, will best ensure the beneficent ends of its institution.

RESOLVED by the Senate and House of Representatives of the United States of America, in Congress assembled, two thirds of both Houses concurring, that the following Articles be proposed to the Legislatures of the several States, as amendments to the Constitution of the United States, all, or any of which Articles, when ratified by three fourths of the said Legislatures, to be valid to all intents and purposes, as part of the said Constitution; viz.

ARTICLES in addition to, and Amendment of the Constitution of the United States of America, proposed by Congress, and ratified by the Legislatures of the several States, pursuant to the fifth Article of the original Constitution.

Amendment I
Congress shall make no law respecting an establishment of religion, or prohibit-
ing the free exercise thereof; or abridging the freedom of speech, or of the press; or the right of the people peaceably to assemble, and to petition the Government for a redress of grievances.

Amendment II
A well regulated Militia, being necessary to the security of a free State, the right of the people to keep and bear Arms, shall not be infringed.

Amendment III
No Soldier shall, in time of peace be quartered in any house, without the consent of the Owner, nor in time of war, but in a manner to be prescribed by law.

Amendment IV
The right of the people to be secure in their persons, houses, papers, and effects, against unreasonable searches and seizures, shall not be violated, and no Warrants shall issue, but upon probable cause, supported by Oath or affirmation, and particularly describing the place to be searched, and the persons or things to be seized.

Amendment V
No person shall be held to answer for a capital, or otherwise infamous crime, unless on a presentment or indictment of a Grand Jury, except in cases arising in the land or naval forces, or in the Militia, when in actual service in time of War or public danger; nor shall any person be subject for the same offence to be twice put in jeopardy of life or limb; nor shall be compelled in any criminal case to be a witness against himself, nor be deprived of life, liberty, or property, without due process of law; nor shall private property be taken for public use, without just compensation.

Amendment VI
In all criminal prosecutions, the accused shall enjoy the right to a speedy and public trial, by an impartial jury of the State and district wherein the crime shall have been committed, which district shall have been previously ascertained by law, and to be informed of the nature and cause of the accusation; to be confronted with the witnesses against him; to have compulsory process for obtaining witnesses in his favor, and to have the Assistance of Counsel for his defence.

Amendment VII
In Suits at common law, where the value in controversy shall exceed twenty dollars, the right of trial by jury shall be preserved, and no fact tried by a jury, shall be otherwise re-examined in any Court of the United States, than according to the rules of the common law.

Amendment VIII
Excessive bail shall not be required, nor excessive fines imposed, nor cruel and unusual punishments inflicted.

Amendment IX
The enumeration in the Constitution, of certain rights, shall not be construed to deny or disparage others retained by the people.

Amendment X
The powers not delegated to the United States by the Constitution, nor prohibited by it to the States, are reserved to the States respectively, or to the people.

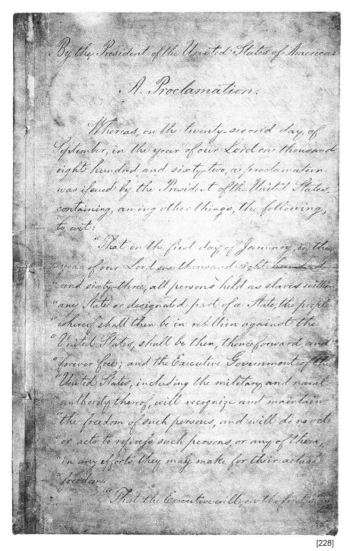

[228]

January 1, 1863
By the President of the
United States of America:

A Proclamation.

Whereas, on the twenty-second day of September, in the year of our Lord one thousand eight hundred and sixty-two, a proclamation was issued by the President of the United States, containing, among other things, the following, to wit:

"That on the first day of January, in the year of our Lord one thousand eight hundred and sixty-three, all persons held as slaves within any State or designated part of a State, the people whereof shall then be in rebellion against the United States, shall be then, thenceforward, and forever free; and the Executive Government of the United States, including the military and naval authority thereof, will recognize and maintain the freedom of such persons, and will do no act or acts to repress such persons, or any of them, in any efforts they may make for their actual freedom.

"That the Executive will, on the first day of January aforesaid, by proclamation, designate the States and parts of States, if any, in which the people thereof, respectively, shall then be in rebellion against the United States; and the fact that any State, or the people thereof, shall on that day be, in good faith, represented in the Congress of the United States by members chosen thereto at elections wherein a majority of the qualified voters of such State shall have participated, shall, in the absence of strong countervailing testimony, be deemed conclusive evidence that such State, and the people thereof, are not then in rebellion against the United States."

Now, therefore I, Abraham Lincoln, President of the United States, by virtue of the power in me vested as Commander-in-Chief, of the Army and Navy of the United States in time of actual armed rebellion against the authority and government of the United States, and as a fit and necessary war measure for suppressing said rebellion, do, on this first day of January, in the year of our Lord one thousand eight hundred and sixty-three, and in accordance with my purpose so to do publicly proclaimed for the full period of one hundred days, from the day first above mentioned, order and designate as the States and parts of States wherein the people thereof respectively, are this day in rebellion against the United States, the following, to wit: Arkansas, Texas, Louisiana, (except the Parishes of St. Bernard, Plaquemines, Jefferson, St. John, St. Charles, St. James Ascension, Assumption, Terrebonne, Lafourche, St. Mary, St. Martin, and Orleans, including the City of New Orleans) Mississippi, Alabama, Florida, Georgia, South Carolina, North Carolina, and Virginia, (except the forty-eight counties designated as West Virginia, and also the counties of Berkley, Accomac, Northampton, Elizabeth City, York, Princess Ann, and Norfolk, including the cities of Norfolk and Portsmouth[)], and which excepted parts, are for the present, left precisely as if this proclamation were not issued.

And by virtue of the power, and for the purpose aforesaid, I do order and declare that all persons held as slaves within said designated States, and parts of States, are, and henceforward shall be free; and that the Executive government of the United States, including the military and naval authorities thereof, will recognize and maintain the freedom of said persons.

And I hereby enjoin upon the people so declared to be free to abstain from all violence, unless in necessary self-defence; and I recommend to them that, in all cases when allowed, they labor faithfully for reasonable wages.

And I further declare and make known, that such persons of suitable condition, will be received into the armed service of the United States to garrison forts, positions, stations, and other places, and to man vessels of all sorts in said service.

And upon this act, sincerely believed to be an act of justice, warranted by the Constitution, upon military necessity, I invoke the considerate judgment of mankind, and the gracious favor of Almighty God.

In witness whereof, I have hereunto set my hand and caused the seal of the United States to be affixed.

Done at the City of Washington, this first day of January, in the year of our Lord one thousand eight hundred and sixty three, and of the Independence of the United States of America the eighty-seventh.

By the President: ABRAHAM LINCOLN
WILLIAM H. SEWARD, Secretary of State.

Gettysburg Address

November 19, 1863

Four score and seven years ago our fathers brought forth on this continent, a new nation, conceived in Liberty, and dedicated to the proposition that all men are created equal.

Now we are engaged in a great civil war, testing whether that nation, or any nation so conceived and so dedicated, can long endure. We are met on a great battle-field of that war. We have come to dedicate a portion of that field, as a final resting place for those who here gave their lives that that nation might live. It is altogether fitting and proper that we should do this.

But, in a larger sense, we can not dedicate -- we can not consecrate -- we can not hallow -- this ground. The brave men, living and dead, who struggled here, have consecrated it, far above our poor power to add or detract. The world will little note, nor long remember what we say here, but it can never forget what they did here. It is for us the living, rather, to be dedicated here to the unfinished work which they who fought here have thus far so nobly advanced. It is rather for us to be here dedicated to the great task remaining before us - that from these honored dead we take increased devotion to that cause for which they gave the last full measure of devotion - that we here highly resolve that these dead shall not have died in vain - that this nation, under God, shall have a new birth of freedom - and that government of the people, by the people, for the people, shall not perish from the earth.

Abraham Lincoln

Civil War Battles in Tennessee

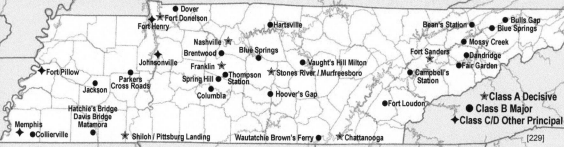

Dover
★ Fort Donelson
Fort Henry
● Hartsville
Bean's Station ●
● Bulls Gap
Blue Springs
Nashville ★
Brentwood ●
Blue Springs
Mossy Creek
Fort Sanders
★
● Dandridge
Johnsonville
Franklin ★
Vaught's Hill Milton
● Fair Garden
✦ Fort Pillow
Spring Hill ●
Thompson Station
★ Stones River / Murfreesboro
Campbell's Station
Parkers Cross Roads
Jackson
Columbia
● Hoover's Gap
● Fort Loudon
Hatchie's Bridge
Davis Bridge
Matamora
Memphis
✦ ● Collierville
★ Shiloh / Pittsburg Landing
Wautatchie Brown's Ferry ●
★ Chattanooga

★ Class A Decisive
● Class B Major
✦ Class C/D Other Principal

[229]

1862	**1863**	**1864**
February 6 **Fort Henry**	*February 2* **Dover/Fort Donelson**	*January 17* **Dandridge**
February 11-16 **Fort Donelson**	*March 5* **Thompson Station**	*January 27* **Fair Garden**
April 6-7 **Shiloh/Pittsburg Landing**	*March 20* **Vaughts Hill / Milton**	*April 12* **Fort Pillow**
May 1862 **Fort Pillow Naval Engagement**	*March 25* **Brentwood**	*August 21* **Memphis**
June 6 **Memphis**	*April 10* **Franklin**	*November 4-5* **Johnsonville**
June 7-8 **Chattanooga**	*June 24-26* **Hoover's Gap**	*Novmber 24-29* **Columbia**
July 13 **Murfreesboro**	*August 21* **Chattanooga**	*November 29* **Spring Hill**
October 5 **Hatchie's Bridge/Davis Bridge/Matamora**	*September 22* **Blountsville**	*November 30* **Franklin**
December 7 **Hartsville**	*October 10* **Blue Springs**	*December 5-7* **Murfreesboro / Wilkinson Pike / Cedars**
December 19 **Jackson**	*October 28-29* **Wauhatchie / Brown's Ferry**	*December 15-16* **Nashville**
December 31 - January 2 1863 **Stones River/Murfreesboro**	*November 3* **Collierville**	
December 31 **Parker's Cross Roads**	*November 16* **Campbell Station**	
	November 23-25 **Chattanooga**	
	November 29 **Fort Sanders / Fort Loudon**	
	December 14 **Bean Station**	
	December 29 **Mossy Creek**	

From Salem to Nashville *Old Glory*

States Admitted to the Union

Rank	State	Admission Date
1	Delaware	December 7, 1787
2	Pennsylvania	December 12, 1787
3	New Jersey	December 18, 1787
4	Georgia	January 2, 1788
5	Connecticut	January 9, 1788
6	Massachusetts	February 6, 1788
7	Maryland	April 28, 1788
8	South Carolina	May 23, 1788
9	New Hampshire	June 21, 1788
10	Virginia	June 25, 1788
11	New York	July 26, 1788
12	North Carolina	November 21, 1789
13	Rhode Island	May 29, 1790
14	Vermont	March 4, 1791
15	Kentucky	June 1, 1792
16	Tennessee	June 1, 1796
17	Ohio	March 1, 1803
18	Louisiana	April 30, 1812
19	Indiana	December 11, 1816
20	Mississippi	December 10, 1817
21	Illinois	December 3, 1818
22	Alabama	December 14, 1819
23	Maine	March 15, 1820
24	Missouri	August 10, 1821

States entered the Union and the great American flag grows to 24 stars [230]

Driver - Benz Family Tree Timeline

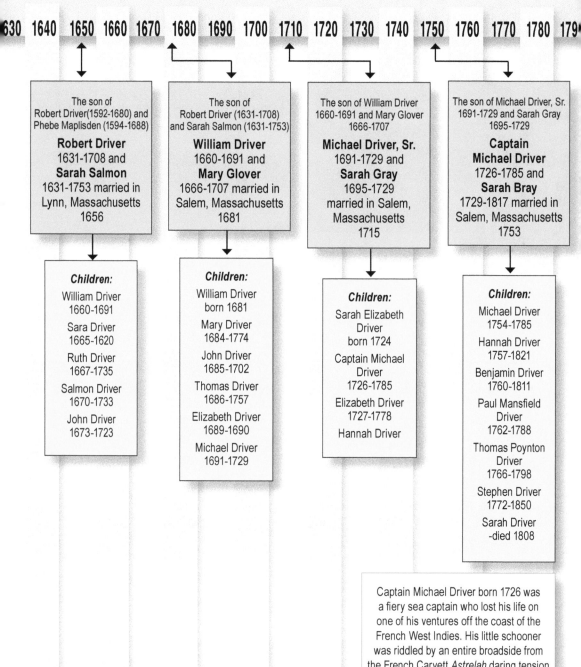

| 630 | 1640 | 1650 | 1660 | 1670 | 1680 | 1690 | 1700 | 1710 | 1720 | 1730 | 1740 | 1750 | 1760 | 1770 | 1780 | 179 |

The son of
Robert Driver(1592-1680) and
Phebe Maplisden (1594-1688)

Robert Driver
1631-1708 and
Sarah Salmon
1631-1753 married in
Lynn, Massachusetts
1656

The son of
Robert Driver (1631-1708)
and Sarah Salmon (1631-1753)

William Driver
1660-1691 and
Mary Glover
1666-1707 married in
Salem, Massachusetts
1681

The son of William Driver
1660-1691 and Mary Glover
1666-1707

Michael Driver, Sr.
1691-1729 and
Sarah Gray
1695-1729
married in Salem,
Massachusetts
1715

The son of Michael Driver, Sr.
1691-1729 and Sarah Gray
1695-1729

Captain
Michael Driver
1726-1785 and
Sarah Bray
1729-1817 married in
Salem, Massachusetts
1753

Children:

William Driver
1660-1691

Sara Driver
1665-1620

Ruth Driver
1667-1735

Salmon Driver
1670-1733

John Driver
1673-1723

Children:

William Driver
born 1681

Mary Driver
1684-1774

John Driver
1685-1702

Thomas Driver
1686-1757

Elizabeth Driver
1689-1690

Michael Driver
1691-1729

Children:

Sarah Elizabeth
Driver
born 1724

Captain Michael
Driver
1726-1785

Elizabeth Driver
1727-1778

Hannah Driver

Children:

Michael Driver
1754-1785

Hannah Driver
1757-1821

Benjamin Driver
1760-1811

Paul Mansfield
Driver
1762-1788

Thomas Poynton
Driver
1766-1798

Stephen Driver
1772-1850

Sarah Driver
-died 1808

Captain Michael Driver born 1726 was
a fiery sea captain who lost his life on
one of his ventures off the coast of the
French West Indies. His little schooner
was riddled by an entire broadside from
the French Carvett *Astrelah* daring tension
between the Americans and the French.
He was thrown into prision at Pointe Petre,
Guadalupe where he died in 1785.

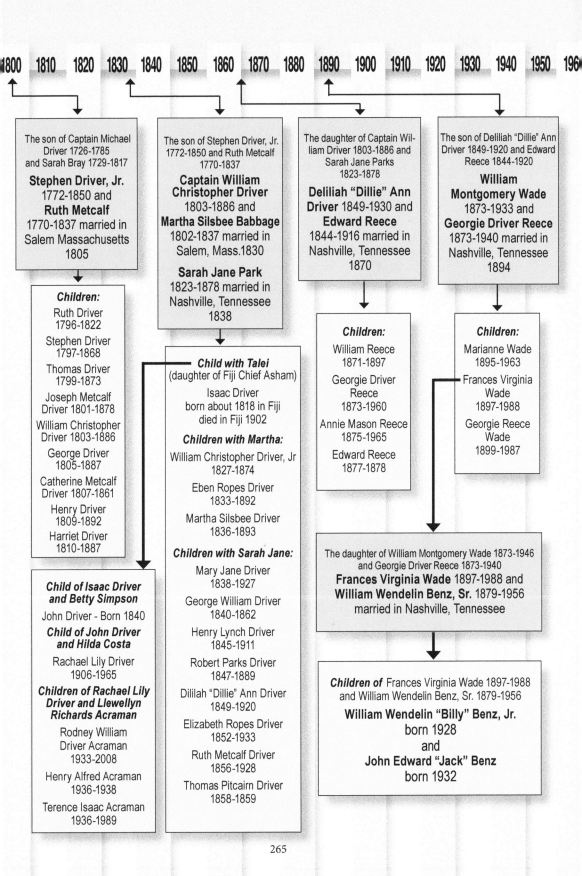

Timeline markers: 1800 | 1810 | 1820 | 1830 | 1840 | 1850 | 1860 | 1870 | 1880 | 1890 | 1900 | 1910 | 1920 | 1930 | 1940 | 1950 | 196•

The son of Captain Michael Driver 1726-1785 and Sarah Bray 1729-1817

Stephen Driver, Jr. 1772-1850 and **Ruth Metcalf** 1770-1837 married in Salem Massachusetts 1805

The son of Stephen Driver, Jr. 1772-1850 and Ruth Metcalf 1770-1837

Captain William Christopher Driver 1803-1886 and **Martha Silsbee Babbage** 1802-1837 married in Salem, Mass.1830

Sarah Jane Park 1823-1878 married in Nashville, Tennessee 1838

The daughter of Captain William Driver 1803-1886 and Sarah Jane Parks 1823-1878

Deliliah "Dillie" Ann Driver 1849-1930 and **Edward Reece** 1844-1916 married in Nashville, Tennessee 1870

The son of Deliliah "Dillie" Ann Driver 1849-1920 and Edward Reece 1844-1920

William Montgomery Wade 1873-1933 and **Georgie Driver Reece** 1873-1940 married in Nashville, Tennessee 1894

Children:
Ruth Driver 1796-1822
Stephen Driver 1797-1868
Thomas Driver 1799-1873
Joseph Metcalf Driver 1801-1878
William Christopher Driver 1803-1886
George Driver 1805-1887
Catherine Metcalf Driver 1807-1861
Henry Driver 1809-1892
Harriet Driver 1810-1887

Child with Talei
(daughter of Fiji Chief Asham)
Isaac Driver
born about 1818 in Fiji
died in Fiji 1902

Children with Martha:
William Christopher Driver, Jr 1827-1874
Eben Ropes Driver 1833-1892
Martha Silsbee Driver 1836-1893

Children with Sarah Jane:
Mary Jane Driver 1838-1927
George William Driver 1840-1862
Henry Lynch Driver 1845-1911
Robert Parks Driver 1847-1889
Dililah "Dillie" Ann Driver 1849-1920
Elizabeth Ropes Driver 1852-1933
Ruth Metcalf Driver 1856-1928
Thomas Pitcairn Driver 1858-1859

Children:
William Reece 1871-1897
Georgie Driver Reece 1873-1960
Annie Mason Reece 1875-1965
Edward Reece 1877-1878

Children:
Marianne Wade 1895-1963
Frances Virginia Wade 1897-1988
Georgie Reece Wade 1899-1987

Child of Isaac Driver and Betty Simpson
John Driver - Born 1840
Child of John Driver and Hilda Costa
Rachael Lily Driver 1906-1965
Children of Rachael Lily Driver and Llewellyn Richards Acraman
Rodney William Driver Acraman 1933-2008
Henry Alfred Acraman 1936-1938
Terence Isaac Acraman 1936-1989

The daughter of William Montgomery Wade 1873-1946 and Georgie Driver Reece 1873-1940
Frances Virginia Wade 1897-1988 and **William Wendelin Benz, Sr.** 1879-1956 married in Nashville, Tennessee

Children of Frances Virginia Wade 1897-1988 and William Wendelin Benz, Sr. 1879-1956
William Wendelin "Billy" Benz, Jr. born 1928 and **John Edward "Jack" Benz** born 1932

Glossary of Nautical Terms

TERM	MEANING
Abeam	Right angle to ship
Aft	To the rear of the ship
Beam	Width of the ship
Bilge	Below the deck
Boom	Spar to which mainsail is attached
Bow	Front of ship
Brig	Two masted sailing vessel
Chronometer	Marine timepiece
Close Haul	Closest sailing point to the wind
Conpanionway	Entrance to below deck
Course	Direction ship is headed
Fathom	Six feet
Fore	To the front of the ship
Foremast	Front mast of the ship
Foresail	Sail in the front of the ship
Gunwale	Side rail of ship
Halyard	Line that hoists a sail
Hands	Ships' crew
Helmsman	Member of crew who steers the ship
Hold	Space below deck for cargo
Hull	Body of ship
Jib	Front sail on ship
Knot	1.15 miles per hour
Lee or Leward Wheel	Steering wheel away from wind
Luff	Back edge of sail
Master	Captain of ship
Mate	Ship regular crew
Rigging, Shrouds	Wires or ropes holding mast in place
Running free	Wind pushing ship from behind
Sextant	Navigation instrument
Sheet	Ropes or lines that control sails
Tack	To change direction to the wind
Topsail	Square sail above the mainsail
Vessel	Ship
Weather Wheel	Steering wheel on windward side of ship
Windward	Toward the wind

I pledge allegiance
to the Flag of the
United States of America,
and to the Republic
for which it stands,
one Nation under God,
indivisible, with liberty
and justice for all.

Folding the Flag

[231]

1. Hold the flag waist-high
with a partner; the flag should be parallel with the ground.

2. Bring the upper and lower halves of the flag together, folding
it lengthwise in half.

3. Fold the flag lengthwise again, bringing the lower half up to
the top. The field of stars should be visible on the left side.

4. Bring the striped corner of the folded edge up to meet the
top edge of the flag, making a small triangle. Fold the triangle
over itself, making the triangle point inward. Continue triangular
folding until the entire length of the flag is folded.

5. When the flag is completely folded, only a triangular blue field
of stars should be visible. Tuck the remaining rectangle into the
triangle's folds.

Old Glory Facts

Historians disagree on the date Driver gave the American flag the name *Old Glory*. Some are of the opinion that he named it on the eve of his voyage to New Zealand on January 14, 1831. Others feel that the correct date was his 21st birthday on March 18, 1824. Most recent historians have chosen the later date as he constantly mentions that *Old Glory* has guided and protected his in all of his travels around the world. We have chosen the date of his birthday as it seems to fit his comments in his later years about the flag. The authors of this book feel that the important fact is not when he named the flag but that he did name the flag—*Old Glory*.

The original flag had only 24 stars. Driver's wife, Sarah Jane, and one of his daughters made repairs to the original flag in 1861, at his request, and sewed on 34 stars to replace the worn originals which, of course, meant that the new stars were in better shape than the rest of the flag. **The anchor was a new one that had not been part of the original flag.**

[233]

Captain Driver's "Old Glory" was originally 24 feet by 12 feet.

Old Glory is at the Smithsonian "where it will always stay" but in the Summer of 2006, it returned for a final eight-month visit to the Tennessee State Museum in Nashville.

Old Glory returns here for eight-month stay at Tennessee State Museum

By ALAN BOSTICK
Staff Writer

The celebrated U.S. flag known as Old Glory has been around the world atop an early 19th-century ship concealed inside a quilt in Nashville during the Civil War, and finally donated to the Smithsonian Institution.

In mid-March, Old Glory returns to Nashville for a special eight month showing at Tennessee State Museum downtown.

"Old Glory: An American Treasure Comes Home" showcases one of two American flags especially rich with history. The other — also owned by the Smithsonian's National Museum of American History but not included in this show — is the Star-Spangled Banner, which inspired Francis Scott Key to pen our national anthem.

"These are both historic flags of a heroic nation," said Lois Riggins-Ezzell, executive director of the museum. "They go hand-in-hand as relics of our heroic past."

Measuring 10-feet-by-17-feet,

Old Glory was presented to a Massachusetts sea captain named William Driver by his family before he set sail around the world in the early 1820s.

The story goes that he exclaimed, "Old glory!" upon first seeing it.

At one point, Driver's ship, the *Charles Doggett*, rescued survivors of the infamous mutiny on the *Bounty* in the South Pacific.

Driver retired to Nashville in the late 1830s and lived on Fifth Avenue South, Riggins-Ezzell said.

When the Civil War heated up, and Old Glory was in danger of being found and burned, Driver, a staunch Unionist, is said to have had it sewn into his bed covers. Later it flew briefly from the state Capitol and was eventually given to the Smithsonian in 1922. Driver is buried in the Nashville City Cemetery.

The state museum show also will feature other historic flags and replicas of famous U.S. flags. ■

Alan Bostick can be reached at 259-8038 or at abostick@tennessean.com.

If you go
▶ "Old Glory: An American Treasure Comes Home," Tennessee State Museum, Fifth Avenue and Deaderick Street, March 17-Nov. 26. Free. Call 741-2692.

A true world traveler, the historic 19th-century flag Old Glory returns to Nashville next month for a special showing.

[234]

America The Beautiful

O beautiful for spacious skies, For amber waves of grain,
For purple mountain majesties Above the fruited plain!
America! America! God shed His grace on thee
and crown thy good with brotherhood
From sea to shining sea!

O beautiful for pilgrim feet,
Whose stern, impassioned stress
A thoroughfare for freedom beat
Across the wilderness!
America! America!
God mend thine every flaw,
Confirm thy soul in self-control,
Thy liberty in law!

O beautiful for heroes proved
In liberating strife,
Who more than self their country loved
And mercy more than life!
America! America!
May God thy gold refine,
Till all success be nobleness,
And every gain divine!

O beautiful for patriot dream
That sees beyond the years
Thine alabaster cities gleam
Undimmed by human tears!
America! America!
God shed His grace on thee
And crown thy good with brotherhood
From sea to shining sea!

[246]

"Our Emblems of Liberty" [235]

COPYRIGHTED 1908 BY JULIUS BIEN & CO N.Y.

Before March 3, 1931, when an act of Congress replaced
"America The Beautiful" with "The Star-Spangled Banner,"
it was our national anthem sung to the tune of "God Save the King."

The Star-Spangled Banner

September 13, 1814 Francis Scott Key, a 34 year old lawyer, left Baltimore in a flag of truce for the purpose of getting a friend released from the British fleet, who had been captured at Marlborough. He went as far as the mouth of the Patuxent River, and was not permitted to return lest the intended attack on Baltimore should be disclosed. He was therefore brought up the Bay to the mouth of the Patapsco, where the flag vessel was kept under the guns of a frigate. He watched the bombardment of Fort McHenry and the flag at the Fort throughout the day with an anxiety that is felt better than described. In the night he watched the bomb shells, and at early dawn he saw the proudly waving flag of his country. He put his thoughts on paper while still on board the ship. His brother-in-law, commander of a militia at Fort McHenry, read Key's work and had it distributed under the name "Defence of Fort M'Henry." The Baltimore Patriot newspaper soon printed it, and within weeks, Key's poem, now called "The Star-Spangled Banner," appeared in print across the country, immortalizing his words.

Oh, say can you see, by the dawn's early light,
What so proudly we hailed at the twilight's last gleaming?
Whose broad stripes and bright stars, through the perilous fight,
O'er the ramparts we watched, were so gallantly streaming?
And the rockets' red glare, the bombs bursting in air,
Gave proof through the night that our flag was still there.
O say, does that star-spangled banner yet wave
O'er the land of the free and the home of the brave?

On the shore, dimly seen through the mists of the deep,
Where the foe's haughty host in dread silence reposes,
What is that which the breeze, o'er the towering steep,
As it fitfully blows, now conceals, now discloses?
Now it catches the gleam of the morning's first beam,
In full glory reflected now shines on the stream:
'Tis the star-spangled banner! O long may it wave

O'er the land of the free and the home of the brave.
And where is that band who so vauntingly swore
That the havoc of war and the battle's confusion
A home and a country should leave us no more?
Their blood has wiped out their foul footstep's pollution.
No refuge could save the hireling and slave
From the terror of flight, or the gloom of the grave:
And the star-spangled banner in triumph doth wave
O'er the land of the free and the home of the brave.

Oh! thus be it ever, when freemen shall stand
Between their loved homes and the war's desolation!
Blest with victory and peace, may the heaven-rescued land
Praise the Power that hath made and preserved us a nation.
Then conquer we must, for our cause it is just,
And this be our motto: "In God is our trust."
And the star-spangled banner forever shall wave
O'er the land of the free and the home of the brave!

This is Jack Van Hooser introducing the three personalities responsible for publishing the book from which you are reading. It is a coincidence that I know all three in the context of book publishing. I attended college with one and the other two helped facilitate two books for which I was the publisher.

Jack Benz with crewmen Matt Steinhauer and Davis McConnico sailing his San Juan 21 sailboat called *Old Glory* on Percy Priest Lake, 1977.[236]

About the Contributors

Author and Publisher
John E. (Jack) Benz

Jack is a lifelong Tennessean currently living in Goodlettsville, Tennessee along with his wife, the former Noroma Carr of Madison and his Shih Tzu buddy, Bentley. He is the proud father of two daughters, Penney Jones and Kelley Benz, both of Sumner County, Tennessee.

He grew up in the Inglewood section of Nashville and graduated from the legendary Isaac Litton High School, known nationally as a school of distinction in academics, athletics, and band.

Jack attended Tennessee Tech and later graduated from Belmont University in its first graduating class in 1955 with a degree in Business Administration. He is an avid supporter of basketball programs at Belmont where he played his last two years. Presently in retirement, he continues to serve the University as a Trustee Emeritus and is seen regularly at courtside for home and away basketball games.

After serving his commitment in the U. S. Army, Jack entered and completed a 50-year career in the insurance industry.

For over 40 years he was active in sailboat racing, winning many events throughout the eastern waters of the United States. His sailing has also taken him to the great lakes, the Caribbean, and the Atlantic Ocean.

Jack Benz has a winning attitude, infectious personality, and relentless work ethic, the 65 years I have known him. These traits will be obvious to the readers of this spellbinding account of his famous great-great-grandfather. Considering his reputation for having exceptional organizational habits and skills, it is not surprising that he has been collecting resources for this undertaking for over 50 years. I am pleased to introduce Jack Benz as a great American with an all-American story.

—Jack Van Hooser

Co-Author and Editor
Garrett Williams
a.k.a. John "Ed" Garrett

Ed was born and nurtured in Springfield, Tennessee, and has lived most of his adult years in the Madison Community of Metropolitan Nashville and Davidson County, Tennessee. He was a four-decade career educator with the Metro Public School District as a classroom teacher, supervisor and H.R. Director for the Secondary Schools. He married his high school sweetheart, the former Jo Ann Adams and they have two children and six grandchildren.

Ed's talents beyond the classroom have awarded him U.S. patents and copyrights for intellectual property creations including a public television media production, two personal books, and ghost writings resulting in four additional published books and several magazine articles. Ed's pseudonym as a writer is Garrett Williams in honor of his father and mother. I have known him in diverse roles for over 50 years and have utilized his knowledge and skills along with that of the forthcoming Nancy Arnold in the publishing of two books, *A Journey to Remember* and *Unstable Times-Unlikely Outcomes*. With admiration, I call him a "wordsmith." You will enjoy his creative contributions to this book. —Jack Van Hooser

---

You are invited
to visit our website

http://www.captaindriver.com

Graphic Design
and Productioin
Nancy Adams Arnold

Nancy's artistic gifts and creative talents became apparent to all observers during her pre-teen years in Cedar Hill, Tennessee. She utilized nature and country life as her studio for drawing, painting, and Brownie camera photography. At age 12, formal lessons began, followed by teacher recommended summer employment preparing art for silk-screening stadium cushions. Next, were stints with the Newspaper Printing Corporation, Design Graphics, Inc., and the Baptist Sunday School Board facilitating ad and mechanical layouts, as well as preparing camera-ready art.

Her horizons continued to broaden through the years as the printing process changed from hot metal to computer-generated page layout and with a variety of experiences including a trip to Rome and the Holy Land leading to specialization in photography for religious publications. She worked at BSSB, now LifeWay Christian Resources for 17 years fulltime and freelance for a total of 40 years.

Nancy lives in Springfield, Tennessee, with husband Eddie and Lucky, everyone's favorite Chihuahua. She is a proud mom of two boys, grandma to four and great-grandma to three, still enjoying using her skills "to play" as she calls it, in Photoshop and InDesign, retouching old photographs, self-publishing books, designing brochures, genealogical research and other special projects.

Along with this current publication about Captain William Driver, I invite readers to examine her amazing work found in my two aforementioned books where she prepared all of the visuals and layouts. —Jack Van Hooser

Bibliography

Page-Citation

1. 1 https://en.wikipedia.org/wiki/O_Captain!_My_Captain (Accessed Feb. 09, 2019)
3. 2 https://en.wikipedia.org/wiki/Far_Away_Places Joan Whitney and Alex Kramer (Accessed Feb. 09, 2019)
6. 3 Moby Dick - https://www.goodreads.com/quotes/195956-call-me-ishmael-some-years-ago--- never-m (Accessed February 09, 2019)
8. 4 https://www.amazon.com/Misguided-Notions-Resurrection-Public- ly-Education/dp/1450295142-look inside page 2 prologue in paragraph 1 (Accessed February 09, 2019.)
8. 5 https://www.amazon.com/Misguided-Notions-Resurrection-Publicly-Education/ dp/1450295142-look inside page 2 prologue in paragraph 2 (Accessed February 09, 2019.)
12. 6 https://www.bufordajohnsonchapter.org/our-d.o.t.a..html (Accessed February 09, 2019).
12. 7 https://genographic.nationalgeographic.com/land-bridge/ (Accessed February 10, 2019)
12. 8 http://articles.latimes.com/1999/apr/11/news/mn-26401 (Accessed February 10, 2019).
12. 9 https://www.history.com/news/10-things-you-may-not-know-about- christopher-columbus (Accessed February 10, 2019).
13. 10 https://www.britannica.com/biography/Pedro-Menendez-de-Aviles (Accessed February 10, 2019)
13. 11 https://www.poetryfoundation.org/poems/45392/ulysses (Accessed February 10, 2019)
14. 12 https://www.brainyquote.com/quotes/robert_louis_stevenson_133860 (Accessed February 10, 2019).
16. 13 https:// www.plimoth.org/learn/just-kids/homework-help/who-were-pilgrims (Accessed February 15, 2019)
17. 14 Who Were The Pilgrims? | Plimoth Plantation, https:// www.plimoth.org/learn/just-kids/homework-help/who-were-pilgrims (Accessed February 15, 2019).
17. 15 https://en.wikipedia.org/wiki/Virginia_Dare (Accessed February 10, 2019)
19. 16 https://www.history.com/news/neanderthal-tools-workshop-discovery-poland/ (Accessed February 10, 2019)
20. 17 https://historyofmassachusetts.org/salem-witch-trials-victims/ (Accessed February 10, 2019)
22. 18 https://biblehub.com/psalms/107-3.htm (Accessed February 10, 2019)
24. 19 https://www.monticello.org/site/jefferson/louisiana-purchase (Accessed February 10, 2019)
25. 20 https://www.landofthebrave.info/religion-in-the-colonies.htm (Accessed Feb. 10, 2019)
25. 21 https://en.wikipedia.org/wiki/Universalism (Accessed Feb. 10, 2019)
26. 22 https://www.bls.gov/apps/pages/index.jsp?uREC_ID=206116&type=d (Accessed Feb. 10, 2019)
27. 23 https://archive.org/details/earlyhistoryofna00hume/page/n2 (Accessed Feb 10, 2019)
28. 24 https://patch.com/massachusetts/salem/home-schooling-89605577 (Accessed Feb 10, 2019)
28. 25 https://www.plimoth.org/learn/just-kids/homework-help/childs-role (Accessed February 15, 2019)
30. 26 https://en.wikipedia.org/wiki/Quasi-War (Aaccessed February 15, 2019)
34. 27 http://cdaworldhistory.wikidot.com/italy-birthplace-of-the-renaissance (Accessed February 15, 2019)
35. 28 https://dspace.mit.edu/bitstream/handle/1721.1/64052/(Accessed February 10, 2019)
35. 29 https://www.smithsonianmag.com/history/salem-sets-sail-2682502/Doug Stewart html (Accessed Feb. 10, 2019)
35. 30 Jack Benz family documents collection by permission. (Accessed May 5,2019)
37. 31 https://www.bible.com/bible/296/NUM.23.23.GNB (Accessed Feb. 10, 2019)
39. 32 https://en.wikipedia.org/wiki/Joint-stock_company. (Accessed May, 2019)
39. 33 https://en.wikipedia.org/wiki/East_India_Company (Accessed Feb. 10, 2019)
41. 34 https://en.wikipedia.org/wiki/Boston_Tea_Party (Accessed June, 2018) https://www.rd.com/food/fun/why-saffron-worlds-most-expensive-spice/
42. 35 https://en.wikipedia.org/wiki/Intolerable_Acts. (Accessed May 5, 2019)
43. 36 https://www.rd.com/food/fun/why-saffron-worlds-most-expensive-spice/ (Accessed May 5, 2019)
47. 37 https://en.wikipedia.org/wiki/Namaste (Accessed May 5, 2019)
47. 38 https://www.malaysiakini.com/letters/64255 (Accessed May 5, 2019)
48. 39 https://en.wikipedia.org/wiki/Snake_charming (Accessed May 5, 2019)
49. 40 https://imagesofhistory.wordpress.com/about/ Dimitri collection(Accessed February 10, 2019)
50. 41 https://library.timelesstruths.org/music/Rescue_the_Perishing/ (Accessed 20 Feb. 2019)
51. 42 https://ninaalvarez.net/category/invictus/ (Accessed 20 Feb. 2019)
53. 43 https://en.wikipedia.org/wiki/ Jason (Accessed 20 Feb. 2019).
53. 44 http://museums.bristol.gov.uk/details.php?irn=112431(Accessed 20 Feb. 2019)
54. 45 https://www.thoughtco.com/war-of-1812-battle-fort-mchenry-2361371 (Accessed 20 Feb. 2019).
55. 46 https://havanatourcompany.com/morro-castle-cuba/ (Accessed 20 Feb. 2019).
57. 47 Holt's Clubhouse. (2019). History of Cuban Cigars.https://www. holts.com/clubhouse/cuban-cigars/history-of-cu-ban-cigars (Accessed 21 Feb. 2019).
57. 48 https://www.nationsonline.org/oneworld/History/Cuba-history.htm (Accessed 20 Feb. 2019).
59. 49 https:// en.wikipedia.org/wiki/Sailors%27_superstitions (Accessed 21 Feb. 2019).
59. 50 https://www.biblegateway.com/passage/?search=2+Timothy+1%3A7&version (Accessed 21 Feb. 2019).
61. 51 Jennifer Williamson by permission-Aim Happy. (2019). A Short Poem about Creativity to Invite New Energy. https:// aimhappy.com/poem-about-creativity/ (Accessed 21 Feb. 2019).

62. 52 https://www.britannica.com/place/Gibraltar (Accessed 21 Feb. 2019).

63. 53 http://military.wikia.com/wiki/Tariq_ibn_Ziyad (Accessed 21 Feb. 2019).

64. 54 https://en.wikipedia.org/wiki/Timbuktu (Accessed 6 Mar. 2019).

65. 55 https://www.history.com/news/who-was-the-richest-man-in-history- mansa-musa (Accessed 21 Feb. 2019).

66. 56 https://www.poetryfoundation.org/poems/43997/the-rime-of-the-ancient-mariner-text-of-1834 (Accessed 21 February 2019).

67. 57 https://www.history. com/this-day-in-history/battle-of-trafalgar (Accessed 21 Feb. 2019).

69. 58 https://en.wikipedia.org/wiki/Sailing,_Sailing (Accessed February 21, 2019).

70. 59 https://www.goodreads.com/quotes/tag/contentment (Accessed 21 Feb. 2019).

72. 60 https://www.britannica.com/place/Ceuta (Accessed 21 Feb. 2019).

74. 61 https://traveltips.usatoday.com/native-plants-animals-morocco-100787.html (Accessed 21 Feb. 2019).

76. 62 http://www.mythencyclopedia.com/Go-Hi/Hercules.html (Accessed February 23, 2019).

80. 63 https://liverpoolbuiltonslavery.wordpress.com/2017/05/21/liverpool-built-on-slavery/ (Accessed 23 Feb. 2019).

81. 64 https://www.phrases.org.uk/bulletin_board/62/messages/529.html (Accessed 7 Mar. 2019).

82. 65 https:// www.seeker.com/legendary-viking-sunstone-navigation-solved-1765489280.html (Accessed 23 Feb. 2019).

83. 66 https://en.wikipe- dia.org/wiki/Ptolemy (Accessed 23 Feb. 2019).

85. 67 https://www.worldatlas.com/aatlas/imageg.htm (Accessed 23 Feb. 2019).

87. 68 https://www.nps.gov/safr/learn/historyculture/historicbilgepump.htm (Accessed 7 Mar. 2019).

90. 69 https://idioms.thefreedictionary.com/best- laid+plans+of+mice+ and+men+oft+go+astray (Accessed 23 Feb. 2019).

91. 70 https://www. reference.com/art-literature/saying-oh-tangled-weave-first-practice-deceive-mean-ff37d94b08c57b23 (Accessed 23 Feb. 2019).

92. 71 https:// www.goodreads.com/quotes/14196-the-moving-finger-writes-and-having-writ-moves-on-nor (Accessed 23 Feb. 2019).

95. 72 https://scialert.net/fulltextmobile/?doi=jfas.2016.191.205 (Accessed 23 Feb. 2019)

96. 73 https://www.quora.com/Is-Cape-Horn- the-worlds-most-dangerous-area-for-sailing (Accessed 23 Feb. 2019)

97. 74 https://www.britannica.com/topic/Scylla-and-Charybdis (Accessed 23 Feb. 2019)

100. 75 https://en.wikipedia.org/wiki/Celestial_navigation (Accessed 23 Feb. 2019)

102. 76 http://fun-stuff.americablog.com/2013/06/hubble- telescope-deep-field-video.html (Accessed 23 Feb. 2019).

103. 77 https://astrologyking.com/dubhe-star/ (Accessed 23 Feb. 2019).

103. 78 https://www.space.com/29445-southern-cross-constellation-skywatching.html (Accessed 23 Feb. 2019).

104. 79 https://en.wikipedia.org/wiki/Line-crossing_ceremony(Accessed 23 Feb. 2019).

106. 80 http://solarnavigator.net/history/john_harrison.htm (Accessed 23 Feb. 2019).

107. 81 https://philoso- phyterms.com/occams-razor/ (Accessed 23 Feb. 2019).

109. 82 https://thestreetandthecityul.wordpress.com/2016/03/01/march-1-1565-rio-de-janeiro-is-founded/ (Accessed 23 Feb. 2019).

110. 83 https://rove.me/to/rio-de-janeiro/rio-carnival (Accessed 23 Feb. 2019).

111. 84 https://en.wikipe- dia.org/wiki/Beagle Channel (Accessed 23 Feb. 2019).

112. 85 https://magellanproject.org/tag/enrique/ (Accessed 23 Feb. 2019).

113. 86 https://en.wikipedia.org/wiki/Patagon (Accessed 23 Feb. 2019).

113. 87 http://aquinasmenai.catholic.edu.au/assets/files/Newsletter271117--compressed.pdf (Accessed 6 Mar. 2019)

116. 88 https://explorers.com/2011/11/29/i-am-the-albatross-that-waits-for-you-at-end-of-3/ Accessed 23 Feb. 2019).

117. 89 http://www.cultofweird.com/death/udre-udre-fiji-cannibal/ (Accessed 23 Feb. 2019).

123. 90 http://thinkexist.com/quotation/vice_is_a_monster_of_so_frightful_mien-as_to_be/163287.html (Accessed 23 Feb. 2019).

125. 91 https://www.goodreads.com/author/quotes/16288605.Ranata_Suzuki (Accessed 24 Feb. 2019).

127. 92 http://www.city-data.com/forum/ christianity/1182399-Gods-overruling-directive-permissive-will-freewill.html (Accessed 8 Mar. 2019).

135. 93 http://tgacv.cz/ aj/s_aj/esc/Australia/aus_animals.htm (Accessed 24 Feb. 2019).

136. 94 https://en.wikipedia.org/wiki/Convicts_in_Australia (Accessed 24 Feb. 2019).

137. 95 https://www. bigmerino.com.au/history-of-wool/ (Accessed 24 Feb. 2019).

139. 96 https://www.poemhunter.com/poem/locksley-hall/ (Accessed 24 Feb. 2019).

144. 97 https://www.bartleby.com/248/798.htm (Accessed 24 Feb. 2019).

144. 98 https://en.wikipedia.org/wiki/Doxology (Accessed 24 Feb. 2019).

145. 99 https://teara.govt.nz/ en/1966/kororareka (Accessed 24 Feb. 2019).

146. 100 https://www.kingjamesbi- bleonline.org/Matthew-10-14/ (Accessed 24 Feb. 2019).

146. 101 https://en.wikipedia.org/wiki/P%C5%8Dmare_IV (Accessed 24 Feb. 2019).

146. 102 https://en.wikipedia.org/wiki/P%C5%8Dmare_III public domain accessed Accessed February 15, 2019).

149. 103 https://en.wikipedia.org/wiki/HMS_Pandora_(1779) (Accessed 24 Feb. 2019)

150. 104 https://biblehub.com/mark/6-4.htm (Accessed 24 Feb. 2019).

150. 105 https://www.historyrevealed.com/eras/18th-century/after-the-mutiny-captain-blighs-return/(Accessed 24 Feb. 2019).

150. 106 https://www.biblegateway.com/passage/?search=Luke+21&version (Accessed 24 Feb. 2019).

153. 107 https://www.poemhunter.com/poem/queen-mab-part-i/ (Accessed 24 Feb. 2019).

155. 108 http://www.quotationspage.com/quote/28734.html (Accessed 24 Feb. 2019).

155. 109 https://en.wikipedia.org/wiki/History_of_Nashville,_Tennessee (Accessed 24 Feb. 2019).

159. 110 https://www.poemhunter.com/poem/the-house-by-the-side-of-the-road-2/ (Accessed 24 Feb. 2019).

163. 111 https://www.goodreads.com/quotes/94296-vice-is-a-monster-of-so-frightful-mien-as-to (Accessed 24 Feb. 2019).

164. 112 The Prince: Chapter XVII. [online] Available at: http://con- stitution.org/mac/prince17.htm (Accessed 24 Feb. 2019).

165. 113 To Althea, from Prison. [online] To Althea, from Prison. Available at: https:// www.poets.org/poetsorg/poem/al-thea-prison (Accessed 24 Feb. 2019).

166. 114 Goodreads.com. (2019). A quote by Augustine of Hippo. [online] Available at: https://www.goodreads.com/quotes/126110-right-is-right-even-if-no-one-is-doing-it (Accessed 24 Feb. 2019).

166. 115 Thirteen.org. (2019). The Supreme Court. The First Hundred Years. Court History | PBS. [online] https://www.thir-teen.org/wnet/supremecourt/antebellum/history2.html (Accessed 24 Feb. 2019).

168. 116 Kingjamesbibleonline.org. (2019). 2 TIMOTHY 2:15 KJV "Study to shew thyself approved unto God, a workman that needeth not to be ashamed, rightly dividing the word of truth.". [online] Available at: https://www.kingjamesbibleon-line.org/2-Timothy-2-15/ (Accessed 24 Feb. 2019).

168. 117 Kingjamesbibleonline.org. (2019). GENESIS 11:7 KJV "Go to, let us go down, and there confound their language, that they may not understand one another's speech.". [online] Available at: https://www.kingjamesbibleonline.org/Gene-sis-11-7/ (Accessed 24 Feb. 2019).

172. 118 GotQuestions.org. (2019). What is the biblical account of Shem, Ham, and Japheth?. [online] Available at: https://www.gotquestions.org/Shem-Ham-Japheth.html (Accessed 24 Feb. 2019).

173. 119 Biblehub.com. (2019). Ephesians 2:9 not by works, so that no one can boast.. [online] Available at: https://biblehub.com/ephesians/2-9.htm (Accessed 24 Feb. 2019).

174. 120 Piper, J. (2019). Why Did God Create the World?. [online] Desiring God. Available at: https:// www.desiringgod.org/messages/why-did-god-create-the-world (Accessed 24 Feb. 2019).

174. 121 Kingjamesbibleonline.org. HEBREWS 13:2 KJV "Be not forgetful to entertain strangers: for thereby some have enter-tained angels unawares.".[online] Available at: https://www.king- jamesbibleonline.org/Hebrews-13-2/ (Accessed 24 Feb. 2019).

174. 122 Biblehub.com. (2019). John 8:7 When they continued to question Him, He straightened up and said to them, "Let him who is without sin among you be the first to cast a stone at her.". [online] Available at: https://biblehub.com/john/8-7.htm (Accessed May 5, 2019)

175. 123 HISTORY. (2019). Manifest Destiny. [online] Available at: https://www.history.com/topics/ westward-expansion/man-ifest-destiny (Accessed May 5, 2019).

176. 124 BrainyQuote. (2019). Aristotle Quotes. [online] Available at: https://www.brainyquote.com/quotes/aristotle_100762 (Accessed May 5, 2019).

177. 125 Early history of the Nashville public schools : Hume, Leland, 1864- : Free Download, Borrow, and Streaming : Internet Archive. [online] Internet Archive. Available at: https:// archive.org/details/earlyhistoryofna00hume (Accessed May 5, 2019).

179. 126 Theodosius Issued an Edict. [online] Available at: https://www. christianity.com/church/church-history/time-line/301-600/theodosius-issued-an-edict-11629680.html (Accessed 10 Mar. 2019).

186. 127 HISTORY. (2019). Kansas-Nebraska Act. [online] Available at: https://www.history. com/topics/19th-century/kan-sas-nebraska-act (Accessed May 5, 2019).

187. 128 Abrahamlincolnonline.org. (2019). "House Divided" Speech by Abraham Lincoln. [online] Available at: http://www.abrahamlincolnonline.org/lincoln/speeches/house.htm (Accessed May 5, 2019).

188. 129 HISTORY. (2019). Morse Code & the Telegraph. [online] Available at: https://www.history. com/topics/inventions/telegraph (Accessed 10 Mar. 2019).

190. 130 Family documents in possession of Jack Benz. (Accessed May 5, 2019).

191. 131 Appalachianmagazine.com. (2019). Why Tennessee is Called the "Volunteer State" | Appalachian Magazine. [online] Available at: http://appalachianmagazine.com/2016/05/24/why-tennessee-is-called-the-volunteer-state/ (Accessed 10 Mar. 2019).

192. 132 https://www.dictionary.com/browse/ill-wind-that-blows-no-one-any-good--it-s-an-Heywood (Accessed 10 Mar. 2019).

194. 133 Steelmuseum.org. (2019). William Kelly. [online] Available at: http://www.steelmuseum.org/i-s-hall-of-fame/kel-ly_william.cfm (Accessed 24 Feb. 2019).

194. 134 HistoryNet. (2019). Fort Henry | HistoryNet. [online] Available at: https:// www.historynet.com/fort-henry (Accessed 25 Feb. 2019).

195. 135 HISTORY. (2019). Battle of Fort Donelson. [online] Available at: https://www.history.com/topics/american-civil-war/battle-of-fort-donelson (Accessed 25 Feb. 2019).

196. 136 Jack Hinson-Pelicanpub.com. (2019). Pelican Products:by category, HISTORY. [on- line] Available at: https://pelican-pub.com/products.php?cat=3&pg=15 (Accessed 25 Feb. 2019).

197. 137 Rust, R. (2019). Civil War | Tennessee Encyclopedia. [online] Tennessee Encyclopedia. Available at: https://tennes-seeencyclopedia.net/entries/civil-war/ (Accessed 25 Feb. 2019).

199. 138 Graves, John; Northwest Davidson County, 1985 p 84, 85

203. 139 American Battlefield Trust. (2019). Battle of Appomattox Court House Facts & Sum- mary. [online] Available at: https://www.battlefields.org/learn/civil-war/battles/appomattox-court-house (Accessed 25 Feb. 2019).

203. 140 Battle Hymn of the Republic. [online] Available at: https://en.wikipedia. org/wiki/Battle_Hymn_of_the_Republic (Accessed 11 Mar. 2019).

205. 141 [online] Available at: https://answers.yahoo.com/question/index?qid=20090310162024AAMlH67 (Accessed 25 Feb. 2019).

207. 142 En.wikipedia.org. (2019). Ku Klux Klan. [online] Available at: https://en.wikipedia.org/wiki/ Ku_Klux_Klan (Accessed 11 Mar. 2019).

209. 143 En.wikipedia.org. (2019). Jim Crow laws. [online] Available at: https://en.wikipedia.org/wiki/Jim_Crow_laws (Ac- cessed 12 Mar. 2019).

209. 144 Kimmons, J., Kimmons, J. and profile, V. (2019). A Poem for You. [online] Graefenburgumc.blogspot.com. Available at: http://graefenburgumc.blogspot.com/2011/01/poem-for-you.html (Accessed May 5, 2019).

211. 145 Goodreads.com. (2019). A quote by Henry Wadsworth Longfellow. [online] Available at: https://www.goodreads.com/ quotes/255749-though-the-mills-of-god-grind-slowly-yet-they-grind (Accessed 25 Feb. 2019).

215. 146 Forums, W., Discussion, G., Wilma, D., Para, L. and Wilma, D. (2019). The Nashville Experiment...Prostitution. [online] American Civil War Forums. Available at: https://civilwartalk.com/threads/the-nashville-experiment-prosti- tution.125069/ (Accessed 25 Feb. 2019).

216. 147 Historyonthenet.com. (2019). The Civil War Amendments. [online] Available at: https://www.historyonthenet.com/ authentichistory/1865-1897/1-reconstruction/1-johnson/cwamendments.html (Accessed 25 Feb. 2019).

216. 148 Hayes, R., Era, 1., History, U., System, S. and Limited, S. (2019). Compromise of 1877 - End of Reconstruction: US History for Kids after Hayes [online] American-historama.org. Available at: http://www.american-historama. org/1866-1881-reconstruction-era/compromise-of-1877.htm (Accessed 12 Mar. 2019).

221. 149 Goodreads.com. (2019). A quote from The Prophet. [online] Available at: https:// www.goodreads.com/ quotes/7217051-the-deeper-that-sorrow-carves-into-your-being-the-more (Accessed 25 Feb. 2019).

Page-Image

1. [1] Painting by artist, Dick Elliott by permission

6-7. [2] Dover Publications, Inc., 120 Great Maritime Paintings, Maxime, Maufra – Red Sun; n.d. graphic design and permission granted by Nancy Arnold. Red Sun [2]

8. [3] https://www.spacetelescope.org/images/potw1006a/ (Accessed July 2, 2018) {public domain}Credit: ESA/Hub- ble and NASA   Supernova in Galaxy in NGC 3810

9. [4] https://commons.wikimedia.org/wiki/File:De_Windstoot- {public domain} (Accessed July 2, 2018) _A_ship_ in_need_in_a_raging_storm_(Willem_van_de_Velde_II,_1707).jpg{{PD-US}} Source Rijksmuseum, Amster- dam. SK-A-1848A Ship in Need in a Raging Storm

10. [5] William Driver from family portraits

10. [6] Massachusetts and Tennessee State Flags{public domain}(Accessed July 2, 2018)

11. [7] Photo by Garrett Williams used by permission Captain William Driver (inserts Rodney Acraman and William Driver); standing Great-Great-Grandson Jack Benz

13. [8] Bering Strait, https://www.readex.com/blog/russia-connection-historical-proposals-reestab- lish-land-link-across-bering-strait (Accessed July 2, 2018) {public domain}

15. [9] https://commons.wikimedia.org/wiki/File:1869_pilgrims_Plymouth_Massachusetts_engr_byAndrews_ LC_00035u.jpg {{PD-US}} (Accessed July 2, 2018)   Landing of the Pilgrims

16. [10] https://virtualcurationmuseum.files.wordpress.com/2013/10/peace-pipe.jpg {{PD-US}} (Accessed July 2, 2018) Massasoit and Gov. John Carver smoking a peace pipe.

18. [11] https://commons.wikimedia.org/wiki/File:Friendship_asail_(15728504423).jpg{public domain} (Accessed July 2, 2018) Friendship Underway

20. [12] https://upload.wikimedia.org/wikipedia/commons/7/7e/ (Accessed July 2, 2018) {public domain}Witchcraft at Salem Village

21. [13] https://commons.wikimedia.org/wiki/File:Licht_der_Zeevaert,_Frontispiz.jpg{{PD-US}} (Accessed July 2, 2018) The Light of Navigation, Dutch Sailing Handbook 1608

22. [14] https://commons.wikimedia.org/wiki/File:Salem_shipping_colonial_color.jpg (Accessed July 2, 2018){{PD- US}} Salem Harbor

23. [15] https://upload.wikimedia.org/wikipedia/commons/b/b8/Surrender_of_Lord_Cornwallis.jpg{public domain} (Accessed July 2, 2018) Surrender of Lord Cornwallis

25. [16] https://upload.wikimedia.org/wikipedia/commons/5/56/New-France1750.pngAttribution:I, JF Lepage {public domain}(Accessed July 2, 2018) New France

25. [17] https://www.theclio.com/web/entry?id=20333{public domain} (Accessed July 2, 2018)   First Church in Salem Village

26. [18] http://www.benjamin-franklin-history.org/wp-content/uploads/2016/02/Boston-Latin-School-150x150.jpg (Accessed July 2, 2018){public domain}   First Boston Latin School

27. [19] https://commons.wikimedia.org/wiki/File:New_England_primer.PNG{{PD-US}}(Accessed July 2, 2018) Reading Lesson

27. [20] https://commons.wikimedia.org/wiki/File:New-England_Primer (Accessed July 2, 2018) Enlarged_printed_and_sold_by_Benjamin_Franklin.jpg {{PD-US}} New England Primer

29. [21] Collage crafted and permission granted by Nancy Arnold from PD images Sailorboy (Accessed July 2, 2018)

30. [22] https://en.wikipedia.org/wiki/Quasi-War#/media/File:Combat_naval_pendant_la_quasi_guerre.jpg

31. [23] https://commons.wikimedia.org/wiki/File:Saugus_Iron_Mill_-_forge_with_bellows.JPG (Accessed August 2, 2018)

31. [24] http://www.loc.gov/pictures/resource/nclc.00291 Child Milking a Cow / {PD-US}((Accessed July 2, 2018)

32. [25] Dover Publications, Inc., 120 Great Maritime Paintings, Eugene Boudin – Quay at Honfleur;1865 - graphic design and permission granted by Nancy Arnold. Quay at Honfleur {PD-US}((Accessed July 2, 2018)

34. [26] https://upload.wikimedia.org/wikipedia/commons/thumb/7/7e/Map_of_Italy_%281494%29-en.svg/842px-Map_of_Italy_%281494%29-en.svg.png {public domain} (Accessed July 2, 2018) Livorno, Italy

36. [27] https://commons.wikimedia.org/wiki/File:Bathing_in_Hooghly_River,_Calcutta_(8136082321).jpg (Accessed July 2, 2018)Bathing in River Hoogly https://commons.wikimedia.org/wiki/File:Fort_William,_Calcutta,_1735.jpg (Accessed July 2, 2018){{PD-US}}

38. [28] Fort William https://commons.wikimedia.org/wiki/File:Aernout_Smit_Table_Bay,_1683 _William_Fehr_Collection_Cape_Town.jpg (Accessed July 2, 2018){{PD-US}}

40. [29] Dutch East India Trading Ship https://slavery.iziko.org.za/inhabitantsofthelodge(Accessed July 2, 2018) {public domain}

40. [30] http://www.eastindiacompany.amdigital.co.uk/Introduction/NatureAndScope(Accessed July 2, 2018) {public domain} British East India Charter

41. [31] https://www.rmfa92.org/history/madras-army-east-india-company{public domain} (Accessed July 2, 2018)/ East India Military

42. [32] https://commons.wikimedia.org/wiki/File:Boston_Tea_Party_Currier_colored.jpg(Accessed July 2, 2018){{PD-US}} Boston Tea Party

43. [33] https://commons.wikimedia.org/wiki/File:Return_visit_of_the_Viceroy_to_the_Maharaja_of_Cashmere.jpg (Accessed July 2, 2018){{PD-US}}Return Visit of the Viceroy

44. [34] https://upload.wikimedia.org/wikipedia/commons/f/f4/Crocus_sativus_01_by_Line1.JPG

44. [35] Saffron graphic design using PD images-Nancy Arnold by permission. Saffronhttps://en.wikipedia.org/wiki/Saffron-https://upload.wikimedia.org/wikipedia/commons/1/17/Saffron8.jpg-Saffron{public domain} (Accessed July 2, 2018)

45. [36] https://en.wikipedia.org/wiki/Kalighat_Kali Temple, (Accessed July 2, 2018).jpg, {public domain} http://www.victorianweb.org/history/empire/india/30,http://www.victorianweb.org/history/empire/india/30.jpg graphic design using PD images-Nancy Arnold by permission. Houses of worship in India (Accessed July 2, 2018)

46. [37] https://www.oldindianphotos.in/2011/06/spice-market-india-1875.html?m=02018) Anthony Davis (antiqphoto@earthlink.net) (Accessed July 2, 2018) Spice Market

47. [38] https://commons.wikimedia.org/wiki/File:Namaste_IN_india.jpg-authorHirenWOW-{public domain} (Accessed July 2, 2018) Namaste

47. [39] https://commons.wikimedia.org/wiki/File:India_-_Varanasi_buey_in_market_-_1702.jpg(Accessed July 2, 2018) Market

48. [40] https://commons.wikimedia.org/wiki/File:Snake_charmers,_India_LCCN2001705502.jpg(Accessed July 2, 2018) {public domain} Snake Charmers

52. [41] https://commons.wikimedia.org/wiki/File:Yasmina.Bounty.JPG Yasmina Bounty (Accessed July 2, 2018)

53. [42] https://www.britishtars.com/2017/03/a-view-of-ye-jason-privateer-c1760.html{public domain} (Accessed July 2, 2018)

55. [43] https://commons.wikimedia.org/wiki/File:Dominic_Serres_the_Elder_-_The_Capture_of_Havana,_1762-artist Dominic Serres.jpg The Capture of Havana{public domain}

55. [44] https://www.loc.gov/resource/det.4a05178/(Accessed July 2, 2018) {public domain}   Havana Wharf

56. [45] http://www.trabajadores.cu/20140630/jolgorio-entre-los-tabacaleros/(Accessed July 2, 2018) {public domain} Cuban Hand-Rolled Cigar Artemiseña Factory

57. [46] http://www.marriedtoadventure.com/wp-content/uploads/2014/04/DSC_6123small.jpg{public domain} (Accessed July 2, 2018) Water-Powered Rum Distillery

58. [47] https://nationalhumanitiescenter.org/sugar-mill-sea-society-caribbean-history/ {public domain} (Accessed July 2, 2018) William Clark, "Slaves Cutting the Sugar Cane,"

58. [48] https://wikivisually.com/wiki/Roman_Catholic_Archdiocese_of_San_Crist%C3%B3bal_de_la_Habana. {public domain} (Accessed July 2, 2018) author Tony Hisgett Columbus Cathedral

58. [49] http://nzbirdsonline.org.nz/species/light-mantled-sooty-albatross(Accessed July 2, 2018)

59. [50] Using the Magic Triangle for Speed, Distance and Time (Compound Measures). [online] Available at: https://owlcation.com/stem/Using-the-magic-triangle-for-speed-distance-and-time-compound-measures DST art crafted by Nancy Arnold-used by permission-Distance-Speed-Time {public domain} (Accessed July 2, 2018)

60. [51] https://www.app.com/story/sports/outdoors/fishing/hook-line-and-sinker/2014/08/07/ocean-facts-mariners-measure-speed-knots/13742273/{public domain} (Accessed July 2, 2018) courtesy Oceanmotion.org Log-line

62. [52] https://commons.wikimedia.org/wiki/File:Europe_about_1560.jpg author William Shepherd {{PD-US}}Manipulation by Nancy Arnold by permission{public domain} (Accessed July 2, 2018) The Strait of Gibraltar1560

63. [53] https://urbanintellectuals.com/2016/06/02/claim-of-dark-ages-hides-your-history-mighty-warrior-tariq-ibn-ziyad-leads-conquest-of-spain-in-711-ad/{public domain} (Accessed July 2, 2018) Tariq ibn Ziyad

64. [54] https://commons.wikimedia.org/wiki/{public domain} (Accessed July 2, 2018) File:Mansa.Musa.Amir.jpg Mansa Musa

65. [55] https://blackpast.org/gah/musa-mansa -1280-1337{public domain} (Accessed July 2, 2018) Mansa Musa's Domain

66. [56] https://mydailyartdisplay.wordpress.com/tag/rime-of-the-ancient-mariner/{public domain} (Accessed July 2, 2018) Water, Water, Everywhere

69. [57] https://commons.wikimedia.org/wiki/File:Saturday_night_at_sea.jpg{public domain} (Accessed July 2, 2018) {{PD-US}} Saturday Night at Sea

70. [58] https://commons.wikimedia.org/wiki{public domain} (Accessed July 2, 2018) /File:Barbary_Macaques_-_Rock_Apes_of_Gibraltar.jpg author Aquila Gib Barbary Macaques Apes

71. [59] https://upload.wikimedia.org/wikipedia/commons {public domain} (Accessed July 2, 2018) Moorish_tower_en_bec_in_Gibraltar.jpg (block building in distance)author Prioryman Moorish Castle

72. [60] https://es.wikipedia.org/wiki/Fez_(Marruecos)#/media/File:Leather_dyeing_vats_in_Fes.jpg - author NaSz451{public domain} (Accessed July 2, 2018) Leather Dyeing Vats

73. [61] https://commons.wikimedia.org/wiki/File:Goats_in_an_argan_tree.jpg_ author Grand Parc - Bordeaux, France-(Accessed July 2, 2018) Climbing Goats in Argan Tree

74. [62] https://commons.wikimedia.org/wiki/File:Argane_oil_production.jpg_/{public domain} (Accessed July 2, 2018) Author Chrumps Traditional Method of Making Argan Oil

75. [63] https://commons.wikimedia.org/wiki/File:Cork_oak_trunk_section.jpg_author plantsurfer(Accessed July 2, 2018) Cross Section of Cork Oak

75. [64] https://commons.wikimedia.org/wiki/File:Quercus_suber_corc.JPG_author Carsten Niehaus de:Korkeicheor(-Accessed July 2, 2018)

77. [65] https://en.wikipedia.org/wiki/Pillars_of_Hercules#/media/File:Instauratio_Magna.jpg{{PD-US}}(Accessed July 2, 2018) _author Francis Bacon_ Pillars of Hercules

78. [66] Graphic manipulation by Nancy Arnold -_Dover Pub._ 1/{public domain} (Accessed July 2, 2018) 20 Great Maritime Paintings_059 Claude Lorrain Seaport at Sunset

79. [67] Graphic manipulation by Garrett Williams from PD map and used by permission of Williams First Trip to Europe

82. [68] https://upload.wikimedia.org/wikipedia/commons/1/14/Silfurberg.jpg_ {public domain} (Accessed July 2, 2018) author ArniEin Viking Sun Stone

83. [69] https://commons.wikimedia.org/wiki/File:PSM_V78_D326_Ptolemy.png_{public domain} (Accessed July 2, 2018) Ptolemy the Cartographer

83. [70] https://xt8dob.wordpress.com/2012/05/09/setting-circles-3-iphone-360-protractor-likely-success_{public domain} (Accessed July 2, 2018) Degree Wheel

84. [71] https://www.turtlediary.com/quiz/types-of-angles.html_Courtesy of turtlediary_{public domain} (Accessed July 2, 2018) Common Angles

84. [72] https://www.ck12.org/earth-science/seasons/lesson/Seasons-HS-ES/_{public domain} (Accessed July 2, 2018) The Earth's Tilt

84. [73] Graphic manipulation by Garrett Williams from PD image and used by permission of Williams Navigation Chart

88. [74] https://upload.wikimedia.org/wikipedia/commons/b/b9/Garthsnaid_-_SLV_H91.250-933.jpg author Allen Green (Accessed July 2, 2018) {public domain} The Storm

89. [75] https://commons.wikimedia.org/wiki/File:Vasa-bilge_pump-upper_gun_deck-1.jpg_(Accessed July 2, 2018) author Peter Isotalo_Manual Bilge Pump

93. [76] https://en.wikipedia.org/wiki/Belshazzar%27s_Feast_(Rembrandt) (Accessed July 2, 2018) {{PD-US}}Belshazzar's Feast

94. [77] https://upload.wikimedia.org/wikipedia/commons/thumb/e/ed/Moana_Beach.jpg/1280px- Moana_Beach.jpg

94. [78] Wikimedia Commons. [online] Beche-de-mer https://commons.wikimedia.org/wiki/File:Actinopyga_miliaris.JPG [Accessed 7 Mar. 2019] author:Julien Bidet.

95. [79] Graphic manipulation by Garrett Williams from PD map and used by permission. (Accessed July 2, 2018) East and West Routes from Salem to Fiji

96. [80] https://commons.wikimedia.org/wiki/File:The_Wreck_of_the_Birkenhead.jpg_author Charles Dixon(Accessed July 2, 2018) {{PD-US}}Shipwreck off Southern Africa

96. [81] https://commons.wikimedia.org/wiki/File:Ulysses_and_the_Sirens_by_H.J._Draper.jpg (Accessed July 2, 2018) {{PD-US}} Odysseus Being lured by the Sirens

97. [82] http://chase-grace-and-the-wrath-of-apollo.wikia.com/wiki/File:Scylla_and_Charybdis.jpg(Accessed July 2, 2018) _ Scylla_and_Charybdis

98. [83] Graphic manipulation from PD image offered by Garrett Williams Base 10 finger counting system

99. [84] http://margaretmuirauthor.blogspot.com/2012/11/cooking-on-wooden-sailing-ships-in.html_{public domain} (Accessed July 2, 2018) Fireheath - HMS Victory (c1770) - replica - Portsmouth Historic Dockyard_Cooking on Sailing Ships

100. [85] http://youcansurvive.net/wp-content/uploads/2013/02/Shadow_Stick.jpg_{public domain} (Accessed July 2, 2018) Shadow Compass

100. [86] http://www.math.nus.edu.sg/aslaksen/gem-projects/hm/0203-1-10-instruments/cross_staff.htm(Accessed July 2, 2018) Cross Staff

101. [87] https://commons.wikimedia.org/wiki/File:Sextant-PSF.png{public domain} (Accessed July 2, 2018) Sextant

102. [88] http://www.aimage.org/why-is-the-earth-tilted-wikipedia/{public domain} (Accessed July 2, 2018) The Earth's Tilt

102. [89] https://oneminuteastronomer.com/860/measuring-sky(Accessed July 2, 2018) / Sky Ruler

103. [90] https://commons.wikimedia.org/wiki/File:Ursa_Major_-_Ursa_Minor_-_Polaris.jpg Author Bonč (Accessed July 2, 2018) - Ursa Major - Ursa Minor - Polaris.jpg

103. [91] http://nzastronomy.co.nz/pages/the-southern-hemisphere-sky(Accessed July 2, 2018) Southern Cross

104. [92] https://commons.wikimedia.org/wiki/File:Line-crossing_ceremony_aboard_M%C3%A9duse-Jules_de_Caudin-IMG_4783-cropped.JPG_author Jules de Caudin {public domain} (Accessed July 2, 2018) {{PD-US}} Crossing the Line 0° Latitude—The Equator

106. [93] https://commons.wikimedia.org/wiki/File:HMS_Association_(1697).jpg(Accessed July 2, 2018) {{PD-US}} author un   known The Scilly Naval Disaster of 1707— prompted The Longitude Act

107. [94] https://upload.wikimedia.org/wikipedia/commons/1/14/Harrison%27s_Chronometer_H5.JPG Racklever at English Wikipedia/(Accessed July 2, 2018) {public domain} John Harrison Chronometer

108. [95] Graphic manipulation from PD image offered by Garrett Williams World Map

109. [96] https://upload.wikimedia.org/wikipedia/commons/8/86/Nicolas-Antoine_Taunay.jpg Source Museu Nacional de Belas Artes {public domain} (Accessed July 2, 2018) View of the Rio Bay as seen from the convent of Santo Antônio

110. [97] https://commons.wikimedia.org/wiki/File:Lingelbach_Karneval_in_Rom_001c.jpg -(Accessed July 2, 2018) artist Jan Lingelbach Carnival

111. [98] https://upload.wikimedia.org/wikipedia/commons/3/33/Strait_of_Magellan.jpeg(Accessed July 2, 2018) {{PD-US}} Strait_of_Magellan

113. [99] http://reliuresetdorures.blogspot.com/2015/07/published-by-rbh-july-2015.html-(Accessed July 2, 2018) Patagonian Giants

114. [100] https://www.sciencefunk.com/2017/09/why-ocean-water-dont-mix.html(Accessed July 2, 2018) Why Ocean Waters Do Not Mix

115. [101] https://images.search.yahoo.com/ https://sites.google.com/a/lewistonpublicschools.org/earth-systems-science-website---grade-9/_/rsrc/1468861235466/home/1st-semester/plate-tectonics/summative-assessment/mapping-quakes/world4.jpg

115. [102] Phillips Library, Peabody Essex Museum Charles Doggett ship log DST

115. [103] https://commons.wikimedia.org/wiki/File:CapeHorn.jpg - (Accessed July 2, 2018) Originally loaded into en-wiki by Wikipedia:User:Pietbarber, as Wikipedia:Image:P1170019.jpg. Cape Horn

116. [104] https://de.wikipedia.org/wiki/Datei:Gibsons_albatross_flight_2.jpg

117. [105] Graphic manipulation of South Pacific PD image by Garrett Williams used by permission Pacific Basin Islands(Accessed July 2, 2018) {public domain}

118. [106] https://commons.wikimedia.org/wiki/File:Cannibalism_on_Tanna.jpeg_artist Charles E. Frazer-{{PD-US}} (Accessed December 12, 2018)-Cannibal Feast

119. [107] https://commons.wikimedia.org/wiki/File:Girl_plucking_tea.jpg (Accessed December 12, 2018)-from Frederic Courtland Penfield-Tamil Girl Plucking Tea(Accessed July 2, 2018) {public domain}

121. [108] https://www.sciencephoto.com/image/554982/large/C0184576-Club_dance, 2018)_Feejee-SPL.jpg Fiji Club Dance(Accessed July 2, 2018) {public domain}

122. [109] https://en.wikipedia.org/wiki/Fijian_¬author Burton Brothers{{PD-US}}(Accessed December 12, 2018) Fijian Couple

123. [110] https://commons.wikimedia.org/wiki/File:3_Samoan_girls_making_ava_1909.jpg-authorBartlett Tripp (Accessed December 12, 2018) Fiji Female Dress Codes After the First Missionaries arrived.

126. [111] Dover Publications, Inc., 120 Great Maritime Paintings, William Bradford-ships in Boston Harbor at Twilight – graphic design and permission granted by Nancy Arnold. Ships in Boston Harbor at Twilight

127. [112] https://www.smithsonianmag.com/history/how-the-flag-came-to-be-called-old-glory-18396/_(Accessed December 12, 2018)

129. [113] https://https://commons.wikimedia.org/wiki/File:24_Star_US_Flag.svg

130. [114] https://en.wikipedia.org/wiki/England_expects_that_every_man_will_do_his_duty (Accessed December 12,

2018) Popin's Code

133. [115] https://en.wikipedia.org/wiki/Leopard_tortoise author Bernard Dupont (Accessed December 12, 2018) Leopard_tortoise

134. [116] https://rephaim23.files.wordpress.com/2015/06/workfuegain.jpg_(Accessed December 12, 2018) Fuegian Giant

134. [117] https://www.bugbog.com/maps/australasia/pacific_map/(Accessed June 18, 2018) Australasia/Pacific map

135. [118] https://en.wikipedia.org/wiki/Great_Barrier_Reef_(Accessed December 12, 2018) author Toby Hudson Great Barrier Reef

135. [119] https://commons.wikimedia.org/wiki/File:Bradshaw_rock_paintings.jpg authorTimJN1 Cave Art Depicting a Giant (Accessed December 12, 2018)

136. [120] https://upload.wikimedia.org/wikipedia/commons/e/e9/Douglas_T_Kilburn_%27South- east Aboriginal_man _and_two_companions%27_1847_daguerreotype_7.8.jpg author Anne O'hehir (Accessed December 12, 2018) Aboriginal man and two companions

136. [121] https://commons.wikimedia.org/wiki/File:Discovery_at_Deptford.jpg(Accessed December 12, 2018) {{PD-US}}Discovery of Depford

137. [122] https://commons.wikimedia.org/wiki/File:Sheep_eating_grass_edit02.jpg_ attribution_fir0002 | flagstaffotos.com.au (Accessed December 12, 2018) Merino Sheep

138. [123] https://commons.wikimedia.org/wiki/File:Edmund_Blair_Leighton_-_A_Wet_Sunday_Morning.jpg {{PD-US}}artist Edmond Leighton_ A Wet Sunday Morning

141. [124] https://upload.wikimedia.org/wikipedia/commons/e/e8/Brooklyn_Museum_-_After_a_Gale-Wreckers-{{PD- US}} _James_Hamilton-overall.jpg (Accessed December 12, 2018) bequest of Nancy Hay-artist James Hamilton-Shipwrecked

141. [125] http://www.salemweb.com/guide/marine - Google Search. [online] Available at: https://www.google.com/search?q=13+http://www.salemweb.com/guide/marine&rlz=1C1CHBF_enUS838US838&source=lnms&tbm=isch&sa=X&ved=0ahUKEwjF3eSajvPgAhUEP60KHY0IBS8Q_AUIECgD&biw=1184&bih=884#imgrc=3DhZxs0MlarcWM: [Accessed 8 Mar. 2019]. Nancy Arnold Graphics-Seven Ships Leave Salem, Six Sank in the Storm.

142. [126] https://commons.wikimedia.org/wiki/File:George_Romney_William_Shakespeare_-_The_Tempest_Act_I,_Scene_1.jpg_ {{PD- US}} (Accessed December 12, 2018) Shakespeare – The Tempest

143. [127] https://www.google.com/search?q=.+https://upload.wikimedia.org/wikipedia/commons/e/e8/Brooklyn_Museum&rlz=1C1CHBF_enUS838US838&source=lnms&tbm=isch&sa=X&ved=0ahUKEwjbtYzF4_PgAhUFT-6wKHT1wAg4Q_AUIESgE&biw=1184&bih=884#imgrc=CLQtpjtUIiI9ZM: background (Accessed December 12, 2018)

144. [128] https://upload.wikimedia.org/wikipedia/commons/0/0b/New_Zealand_relief_map.jpg(Accessed December 12, 2018) New Zealand Relief Map

145. [129] https://kaprizulka.mediasole.ru/9_krupneyshih_porod_sobak_v_mire(Accessed December 12, 2018)A Charley and his Pony Friend

146. [130] https://en.wikipedia.org/wiki/Tahitians_author Jeyan-(Accessed December 12, 2018) Tahiti Girls

147. [131] https://en.wikipedia.org/wiki/P%C5%8Dmare_IV

148. [132] https://commons.wikimedia.org/wiki/File:Mutiny_HMS_Bounty.jpg- source National Maritime Museum{{PD- US}} (Accessed December 12, 2018) Captain Bligh and his Loyalists

149. [133] https://www.theislandwiki.org/index.php/Pitcarin island.jpg

151. [134] https://spectrummagazine.org/sites/default/files/imagecache/large_article_image/Pitcairn_Bay.jpg (Accessed December 12, 2018) Pitcairn Island

154. [135] https://commons.wikimedia.org/wiki/File:Paracas_National_Reserve._Ica,_Peru.jpg author-masT3rOD (Accessed December 12, 2018) The Tide Rolls In

156. [136] https://www.nashville.gov/Parks-and-Recreation/Historic-Sites/Fort-Nashborough.aspx(Accessed December 12, 2018) Fort Nashborough

158. [137] William Driver Home Nashville from family pictures

161. [138] https://commons.wikimedia.org/wiki/File:Old_nashville¬riverfront.jpg

163. [139] https://commons.wikimedia.org/wiki/File:Henry_P._Moore_(American_Slaves_of_General_Thomas_F._Drayton_-_Google_Art_Project.jpg_(Accessed December 12, 2018) Slaves of General Thomas F. Drayton

164. [140] http://2.bp.blogspot.com/_Z- / DEoN3ydt4/ShgNv_N070I/AAAAAAAAs8evzbrI4GTGg/s1600/Trail+of+Tears.jpg (Accessed December 12, 2018)_The Trail of Tears

165. [141] 1853 stereoscopic image of Nashville Suspension Bridge from stereoscopic image supplied by Garrett Williams_1853 Suspension Bridge – Nashville

166. [142] https://commons.wikimedia.org/wiki/File:Trails_of_Tears_en.png author-www.demis.nl - (Accessed December 12, 2018) Indian Relocation Paths with emphasis on Northern Route through Nashville

167. [143] https://www.christcathedral.org/ -(Accessed December 12, 2018) Christ Church Cathedral—Nashville, TN

168. [144] https://www.google.com/search?rlz=1C1CHBF_enUS772US772&tbm=isch&q=/Sinners_in_the_Hands_of_an_Angry_God_by_Jonathan_Edwards&chips=q:sinners_in_the_hands_of_an_Angry_God-(Accessed December 12, 2018)-Famous Sermon

281

169. [145] http://chqdaily.com/2018/07/gary-moore-to-discuss-history-of-camp-meetings-including-fair-point-for-heri-tage- lecture (Accessed December 12, 2018) - Campground Meeting

170. [146] http://www.civicscope.org/nashville-tn/JohnDillahunty_ (Accessed December 12, 2018) First Baptist Church

170. [147] https://rimisstac.essayblog.photography/c900/(Accessed December 12, 2018) The Hegelian Dialectic

171. [148] https://commons.wikimedia.org/wiki/File:Races2.jpg(Accessed December 12, 2018){PD- US}} Ham, Japheth, Shem (Sons of Noah)

172. [149] https://www.smithsonianmag.com/history/slavery-trail-of-tears-180956968/ (Accessed December 12, 2018) Slavery-Trail-of-Tears

174. [150] http://www.pravoslavie.ru/110185.html- (Accessed December 12, 2018) Christ and the Sinner

175. [151] https://commons.wikimedia.org/wiki/File:American_Progress_(John_Gast_painting).jpg{{PD-US}}_(Ac-cessed December 12, 2018)American Progress

177. [152] Professor Alfred Hume image from Early History of Nashville Public Schools pamphlet in the Garrett Wil-liams personal Library and used by permission.

178. [153] https://commons.wikimedia.org/wiki/File:WTN_EVula_036.jpg (Accessed December 12, 2018)Alfred Hume

180. [154] http://www.civicscope.org/nashville-tn/Relevance1850sNashville-(Accessed December 12, 2018) Hume-Fogg High School

181. [155] https://www.loc.gov/resource/cph.3a38293/(Accessed December 12, 2018) Classroom Management

181. [156] https://upload.wikimedia.org/wikipedia/commons/thumb/5/52/19th_century_classroom%2C_Auck-land_-_0795.jpg/1280px- (Accessed December 12, 2018) 19th Century Classroom

182. [157] http://www.icollector.com/Three-19th-century-children-s-school-books_i11077408(Accessed December 12, 2018) 19th Century School Books

182. [158] http://mythfolklore.net/aesopica/bewick/95.htm-(Accessed December 12, 2018) The Crow and the Pitcher — Aesop

183. [159] https://en.wikipedia.org/wiki/The_Goose_That_Laid_the_Golden_Eggs - (Accessed December 12, 2018) The_Goose_That_Laid_the_Golden_Eggs—Aesop

184. [160] Photo by Nancy Arnold

185. [161] https://upload.wikimedia.org/wikipedia/commons/f/f6/Scourged_back_by_McPherson_%26_Oli-ver%2C_1863%2C_retouched.jpg(Accessed December 12, 2018) Mathew Brady photographer Whipped Peter

186. [162] http://bradfieldhasclass.weebly.com/slave-auction.html.

186. [163] Recruiting poster unknown

187. [164] https://commons.wikimedia.org/wiki/File:Fort_sumter_1861.jpg (Accessed December 12, 2018)Fort Sumter under Confederate control author Alma A. Pelot _Fort Sumter under Confederate Control

188. [165] https://en.wikipedia.org/wiki/Fort_Sumter(Accessed December 12, 2018) Fort Sumter under Union Control

188. [166] https://upload.wikimedia.org/wikipedia/commons/3/38/Charleston_Mercury_Secession_Broadside%2C_1860. jpg author Charleston Mercury(Accessed December 12, 2018) Charleston Mercury on Secession

189. [167] https://commons.wikimedia.org/wiki/File:Map_of_CSA_4.png (Accessed December 12, 2018) Confederate States of America

191. [168] https://www.mysanantonio.com/150years/article/Happy-229th-birthday-Alamo-defender-Davy-Crock-ett-6449024.php#photo-8477555 (Accessed December 12, 2018)no caption needed just the number

192. [169] https://commons.wikimedia.org/wiki/File:Fort_Henry_Campaign.png Attribution: Map by Hal Jespersen, www.posix.com/CW Fort Henry Campaign

194. [170] https://commons.wikimedia.org/wiki/File:Washington_Iron_Works_Franklin_County_Virginia.JPG_author Marmaduke Percy_(Accessed December 12, 2018) Typical Iron Ore Furnace

195. [171] https://www.theclio.com/web/entry?id=35712(Accessed December 12, 2018) Fort Henry Surrendered to the Union

196. [172] https://www.battlefields.org/learn/articles/anatomy-river-defenses(Accessed December 12, 2018) Facing North Downstream from Fort Donelson

197. [173] http://www.rarenewspapers.com/view/201249?imagelist=(Accessed December 12, 2018)From Fort Donelson to Clarksville

197. [174] https://commons.wikimedia.org/wiki/File:USSCarondelet.jpg(Accessed December 12, 2018) The USS Caronde-let arrives in Nashville from Fort Donelson

197. [175] https://civilwartalk.com/threads/hospital-ship-uss-nashville.85263/(Accessed December 12, 2018) Hospital Ship on the Banks of the Cumberland at Nashville

198. [176] https://www.britannica.com/event/Battle-of-Nashville/media/403895/221090(Accessed December 12, 2018) State Capital view from Train Station x

198. [177] https://en.wikipedia.org/wiki/William_%22Bull%22_Nelson(Accessed December 12, 2018) General William Bull Nelson

199. [178] http://www.theswedishtiger.com/941-scotts.html(Accessed December 12, 2018)Union Soldiers Camped on Capitol Grounds

199. [179] https://www.durangotexas.com/eyesontexas/textour/granbury/james.htm_(Accessed December 12, 2018) Jesse and Frank James

200. [180] https://commons.wikimedia.org/wiki/File:Old_nashville_riverfront.jpg(Accessed December 12, 2018)Riverside Commerce Downtown Nashville

202. [181] https://upload.wikimedia.org/wikipedia/commons/4/40/Come_and_Join_Us_Brothers%2C_by_the_Supervisory_Committee_For_Recruiting_Colored_Regiments.jpg(Accessed December 12, 2018)Colored Regiment

202. [182] http://www.bonps.org/photos/nashville-at-war/(Accessed December 12, 2018)The Uncivil War

202. [183] https://upload.wikimedia.org/wikipedia/commons/6/67/Appomattox_courthouse.jpg_{{PD-US}}photographer Timothy O'Sullivan-Appomattox Courthouse

203. [184] https://upload.wikimedia.org/wikipedia/commons/b/bc/General_Robert_E._Lee_surrenders_at_Appomattox_Court_House_1865.jpg artist Thomas Nast(Accessed December 12, 2018) Lee's Surrender

204. [185] https://commons.wikimedia.org/https://upload.wikimedia.org/wikipedia/commons/a/a1/Nashville%2C_Tennessee._Fortified_bridge_over_the_Cumberland_River_LOC_cwpb.02090.tif

205. [186] https://upload.wikimedia.org/wikipedia/commons/0/0c/The_Last_Hours_of_Abraham_Lincoln_by_Alonzo_Chappel%2C_1868.jpg(Accessed December 12, 2018)_artist Alonzo Chappel Lincoln's Final Hours

206. [187] https://commons.wikimedia.org/wiki/File:Engine_Nashville_of_the_Lincoln_funeral_train.jpg_author- Library of Congress(Accessed December 12, 2018)Lincoln's Funeral Train-The Old Nashville

206. [188] https://commons.wikimedia.org/wiki/File:Lincoln_and_Johnsond.jpg (Accessed December 12, 2018) -author Joseph Baker-Johnson and Lincoln-Reconstruction Era

207. [189] https://learninglab.si.edu/collections/triumph-and-tragedy-us-reconstruction-1865-1877/GayeuEN7ve29xR-r0#r/465547(Accessed December 12, 2018)Author Thomas Nast-Johnson's Reconstruction

208. [190] https://commons.wikimedia.org/wiki/File:James-robinson-graves-caning.jpg (Accessed December 12, 2018)

209. [191] https://commons.wikimedia.org/wiki/File:William_Gannaway_Brownlow_2.jpg(Accessed December 12, 2018) William Gannaway Brownlow

209. [192] https://upload.wikimedia.org/wikipedia/commons/b/b2/Minstrel_dancer_1.jpg(Accessed December 12, 2018) Jim Crowe Minstrel Dancer

210. [193] https://www.vcreporter.com/2017/02/celebrating-black-history-month-genealogical-society-hosts-lectures-on-researching-african-american-family-history/(Accessed December 12, 2018) Freedmens Segregated School

210. [194] https://upload.wikimedia.org/wikipedia/commons/4/4a/%22Colored%22_drinking_fountain_from_mid-20th_century_with_african-american_drinking.jpg. author Russell Lee Colored Drinking Fountain mid-20th Century

210. [195] https://commons.wikimedia.org/wiki/File:Cabins_for_Colored.jpg (Accessed December 12, 2018)author Marian Post-Colored Cabins

211. [196] https://blogs.brown.edu/libnews/happy-birthday-frederick-douglass/from Lincoln Broadsides collection (Accessed December 12, 2018) Lincoln-Douglass

211. [197] Image from May 6, 1871 Issue of Harper's Weekly. Interior Page woodblock double cartoon, of that Harper's Weekly issue.

213. [198] https://www.atlasobscura.com/articles/how-civil-war-soldiers-gave-themselves-syphilis-while-trying-to-avoid-smallpox(Accessed December 12, 2018)Photography George Henry Fox Smallpox

213. [199] https://www.cdc.gov/std/syphilis/images/rash-gbr.htm(Accessed December 12, 2018) Syphilis

214. [200] http://nashvillesaloons.weebly.com/5-hells-half-acre.html(Accessed December 12, 2018) Nashville's Hells half Acre – Slums, Saloons and Brothels

214. [201] https://www.jocelyngreen.com/index.php?q=2013/04/03/chief-camp-diseases-of-the-civil-war (Accessed December 12, 2018) Union Nurse Annie Bell in Nashville

214. [202] https://victorianparis.files.wordpress.com/2011/07/prostitutes.jpg(Accessed December 12, 2018)Brothel

215. [203] https://blogs.wnpt.org/mediaupdate/2017/02/06/civil-war-era-prostitution

215. [204] https://www.archives.gov/research/recover/prostitute-(Accessed December 12, 2018)Provost granted License to a prostitute

216. [205] https://opinionator.blogs.nytimes.com/2013/12/05/the-nashville-experiment/ (Accessed December 12, 2018) Hospital for Federal Officers

217. [206] https://upload.wikimedia.org/wikipedia/commons/b/b7/Customs_House_Nashville_Tennessee.jpg (Accessed December 12, 2018) Customs_House_Nashville_Tennessee.jpg

218. [207] https://tnsos.org/tsla/imagesearch/images/1283.jpg (Accessed December 12, 2018)-Captain William Driver

219. [208] https://Fold 3.com, farm theft, page 14 Southern Claims 18711880 copy p. 5 of 49

219. [209] https://Fold 3.com, BuenaVista Ferry_page__6-gs300.jpg

220. [210] https://Fold 3.com,William Driver Payment 5528pdf_(page 11 of 14)

221. [211] https://commons.wikimedia.org/wiki/File:Bounty_II_steering_wheel.JPG

223. [212] https://en.wikipedia.org/wiki/Pitcairn_Islands

225. [213] https://commons.wikimedia.org/wiki/File:Pitcairn_John_Adams_Grab.jpg

229. [214] https://tnsos.org/tsla/imagesearch/images/1283.jpg

231. [215] https://www.smithsonianmag.com/history/how-the-flag-came-to-be-called-old-glory-18396/

233. [216] Photo by Nancy Arnold by permission

234.   [217] https://commons.wikimedia.org/wiki/File:Calcutta_1852.jpg
236.   [218] Rodney Acraman clipping from Nashville Tennessean 22 Sept. 1968 Jack Benz Collection
237.   [219] Rodney Acraman photo from MyHeritage photo 300067jpg
238.   [220] Photo by Nancy Arnold by permission
239.   [220] Photo by Nancy Arnold by permission
240.   [221] https://upload.wikimedia.org/wikipedia/commons/a/a6/%22Good_bye%2C_Dad%2C_I%27m_off_to_fight_
              for_Old_Glory%2C_you_buy_U.S._gov%27t_bonds%22_Third_Liberty_Loan_-_-_Lawrence_Harris_%3B_
              Sackett_%26_Wilhelms_Corp._N.Y._LCCN2002711986.jpg
240.   [222] https://www.aoc.gov/art/other-paintings-and-murals/signing-constitution_artist Howard Christy
242.   [223] https://commons.wikimedia.org/wiki/File:Grace_Church_Newark_plaque.jpg
243.   [224] https://en.wikipedia.org/wiki/Articles_of_Confederation#/media/File:Articles_page1.jpg
244.   [225] http://www.ushistory.org/declaration/document/ Declaration of Independence
248.   [226] https://www.archives.gov/founding-docs/constitution-transcript
256.   [227] https://billofrightsinstitute.org/founding-documents/bill-of-rights/
258-259. [228] https://www.archives.gov/exhibits/featured-documents/emancipation-proclamation
260.   Graphic Collage by Nancy Arnold by permission
262.   [229] Civil War Battles Civil War collage
263.   [230] States Admitted to Union States Collage by Nancy Arnold by permission
264-265.   Driver-Benz family tree graphics by Nancy Arnold by permission
267.   Pledge of Allegiance graphics by Nancy Arnold by permission
268.   [231] http://www.usflag.org/fold.flag.html
269.   [233] Smithsonian Institution, National Museum of American History
269.   [234] Jack Benz Collection
270.   [235] America The Beautiful
271.   [236] http://www.clker.com/clipart-772739.html